MY ANCESTOR WAS A COALMINER

by David Tonks

SOCIETY OF GENEALOGISTS ENTERPRISES LTD

Published by
Society of Genealogists Enterprises Limited
14 Charterhouse Buildings, Goswell Road, London EC1M 7BA
© The Society of Genealogists Enterprises and the author 2010.

ISBN: 978-1-903462-71-3

British Library Cataloguing in Publication Data
A CIP Catalogue record for this book is available from the British Library.

The Society of Genealogists Enterprises Limited is a wholly owned subsidiary of the Society of Genealogists, a registered charity, no 233701.

Dedication

This second edition is dedicated to Amy Anne.

About the Author

David Tonks was brought up in an East Durham colliery village, and later moved to the nearby new town of Peterlee, named after the famous miners' leader. In 1963 he left school to embark on a career in the public library service. He was awarded a Fellowship by the Library Association in 1975 and also holds a Diploma in Management Studies and a certificate in Local History, and is a graduate of the Open University. He has been interested in family history for over forty years.

David married a fellow-librarian in 1969 and has two sons, three grandsons and a granddaughter. He has coalminers on both sides of his family, with one of his proudest possessions being a photograph and newspaper report of the diamond wedding celebrations of his pitman *maternal* great-grandfather in 1947 – he was present as a very small baby! His *paternal* great-great-great grandfather left the South Staffordshire coalfield in the 1830s to move to the expanding North East pits. Coincidentally both families lived at different times in the same small mining community of Brickgarth, Easington Lane, Co. Durham.

After a few years in the Midlands, David has now returned to the North East, where he is continuing his researches into family, local, Methodist, and mining, history.

FOREWORD

Fifty years ago, there were over a million coal miners in Britain. The country was completely reliant on coal for its energy needs, and the men who dragged this 'black gold' from the earth were considered a special breed. They were admired for their courage and physical strength, and feared for their industrial and political strength and for the perceived roughness of their way of life. The miner was a central figure in the national consciousness; the hard man of heavy industry; resilient, proud and grimly humorous; whose work in the most dangerous and uncomfortable of conditions – though undertaken only for his own survival – represented a sort of sacrifice on behalf of the rest of us. This folklore image of the miner remains potent, and it is based in a large measure in reality. Yet for most of us, miners remain an undifferentiated mass. From the hewers of the old pick and shovel days, working virtually naked and covered in black filth, to the power loaders of more recent times, in their fluorescent overalls, for the public at large, a miner is a miner.

But when I began researching my book *Coming Back Brockens*, about the village of Horden, I discovered that every mine was a world, where a vast number of trades were employed. Mechanics, electricians, bricklayers blacksmiths – with all their sub-divisions: fitters, joiners, horseshoers, cage smiths, tub menders – all of whom came broadly within the category of miner. There were the 'back-by men' who maintained the underground roadways, the 'putters' who took the empty tubs to the coal face and brought the 'full uns' out, the 'datal lads' straight out of school who had to do whatever was asked of them. There were the people who looked after the ponies, the men who operated the lifts, the officials – deputies, overmen and under-managers – had all worked their way up though the mining ranks. The names of these different positions varied in different parts of the country – and even from pit to pit – but each branch of work had its identity within a community of labour that might comprise several thousand men.

Among these miners there were radically different kinds of people: miners who were fanatical about sport, miners who lived in their allotments, in their pigeon crees or out in the hedgerows, setting traps on the gaffer's land. There were miners who lived for the brass band, miners who lived to be in the cosy fug of 'the club'. Miners who weren't concerned with much beyond their beer and baccy for the coming week, and those who were active in the union and local politics – who dreamed of 'raising their class' and creating a new world. There were miners who could barely sign their name and miners who wrote books and had their paintings exhibited all over the world. I even came across a miner who wrote an oratorio.

This book will help you discover the world of your mining ancestors – a world that is, very likely, more complex and fascinating than you've ever imagined. And in the process it will help you understand some of the things that have come to make you the person you are.

Mark Hudson

(*Coming Back Brockens* is published by Vintage paperbacks).

CONTENTS

Foreword iii
Illustrations vii
Acknowledgements ix
Note xi

PART I MINERS AND MINING

1. Coalmining in Great Britain 1
Formation 1
Early history 1
Sea Coal 3
Industrial Revolution 3
Transport 3
19th century development 4
World War I 4
Decline 4
World War II 5
Nationalisation 5
Opencast mining 5
Further decline 6
Rundown 6
Privatisation 7
Into the new millennium 7
Online resources 7

2. Why is he different? 9
Housing 12

Lodgers 13
Migration 13
Age at death 16
Church Records 16
Family size 19
In-breeding 19
The Bond 20
Butties and Truck 21
Disaster victims 22
Serfdom 22
Government reports 23

3. Work **25**
The Job 25
Danger 29
Disasters 30
Unions 31
Strikes 33
Banners 34

4. How did he live? **37**
Communities 37
Families and family life 42
Life stories 43

5. Was he church or chapel? **45**
Additional Sources 51

6. How did he spend his leisure time? **53**
Additional Sources 56

7. Was he a Bevin Boy? **57**
Additional Sources 60

8. But my mining ancestor was a woman! **61**
Additional Sources 64

9. The Arts **67**
Art 67
Literature 69
Music and Song 70
Film 70

PART II DOCUMENTARY EVIDENCE   **71**

10. The Elemore Colliery Disaster   **71**
2 December 1886

PART III SOURCES   **85**

11. National Collections   **85**

12. Coalfield Collections   **91**
Northumberland and Durham   92
Cumberland   101
Lancashire and Cheshire   103
Yorkshire   109
The East Midlands   116
The West Midlands   121
Gloucestershire and Somerset   127
Kent   131
Wales   132
Scotland   142

13. Printed and digital sources   **149**
Bibliography   149
General literature: Classics   150
General literature: Modern histories   150
Digital   152
Websites   152
Mailing Lists   154
Web Rings   154
CD ROMs   154

Appendix I   **155**
Glossary   **155**
Notes   **169**
Index   **180**
About the Society of Genealogists   **186**

List of illustrations:
Frontispiece   Coalmining counties of Great Britain.   xii
Plate I   Fifteenth-century Forest of Dean miner.   2
Plate II   Moving house.   14
Plate III   'Barrington' Baptismal Register, Longbenton   18
    1805.

Plate IV   Hewer in a Durham pit about 1900.   27
Plate V   Colliery officials, 1904.   28
Figure 1   Mining disasters with more than 100 fatalities.   31
Plate VI   Banner of the NUM. North East Area.   35
Plate VII   Colliery dwellings.   38
Plate VIII   Mining community, 1920s.   41
Plate IX   Chapel.   47
Plate X   Bevin boys, Cramlington, 1944.   58
Plate XI   Margaret Hipps.   63
Plate XII   Census Enumerator's Book Brickgarth 1881.   73
Plate XIII   Burial Register, St Michael Lyons, December 1886.   75
Plates XIV-   *Durham Chronicle*, 10 December 1886.   76,
XV   77
Plate XVI   *Palmer's Index to 'The Times' Newspaper, Oct 1 to*   78
                Dec 31, 1886.
Plate XVII   *The Times*, 3 December 1886.   79
Plate XVIII   *Reports…on the circumstances attending an explosion…*   80
Plate XIX   *Reports of The Inspectors of Mines … 1886.*   81
Plate XX   Plan shewing the Ventilation of Elemore Colliery.   82
Plate XXI   *Sunderland Daily Echo,* 3 December 1886.   83
Plate XXII   The memorial in the churchyard at St Michael, Lyons.   84

Sources:

Beamish: the North of England Open Air Museum – Plate IV

Bevin Boys Association – Plate X

Census Enumerator's Book, 1881 (courtesy H.M.S.O.) – Plate XII

Children's Employment Commission *Report* (courtesy H.M.S.O.) – Plate XI

Gil Colman – Plate XXII

Robert MacDonald (courtesy Mary Stobbart) - V

Northumberland Record Office (courtesy Rev. P S Ramsden) – Plate III

Parliamentary Papers (courtesy H.M.S.O.) – Plates XVIII-XX

St Michael's Church, Lyons (courtesy Rev. M Beck) – 19

Sunderland Libraries – Plates XIV-XV, XXI

Times Newspapers – Plates XVI-XVII

Tonks Collection – Plates I-II, VII-IX

Trade Union Printing Services (TUPS) – Plate VI

ACKNOWLEDGEMENTS

First Edition

This book could not have been written without the help of many people, to whom my sincere thanks are due.

David Butler (Durham Record Office), Phil Hall (Sunderland Local Studies Centre), Terry Knight (Cornish Studies Library) and Sue Wood (Northumberland Record Office) who advised on the contents of the questionnaires.

The managers of the many collections in archive centres, record offices, libraries and museums who took time to complete and return the questionnaire, and sometimes to answer follow-up queries.

The copyright holders who gave permission to reproduce material.

The staff of Durham City Local Studies Library, Grindon Library, Lanchester Library, Mid-Staffordshire Postgraduate Medical Centre Library, Newcastle City Library, Staffordshire Libraries, Sunderland City Library, Sunderland Schools Library Service; the libraries of the Northumberland and Durham Family History Society, and of the Society of Genealogists, for the support of their staff and the use of their stocks; and access to the Internet. Special thanks go to John Hall and his staff at Durham University Library, without whose wonderful collections this book would never have been written.

The Society of Genealogists for agreeing to my suggestion that this book was needed, and to Else Churchill, Mark Herber and John Titford.

Bill Eckford, for the loan of a PC at a vital moment.

Gil Colman, for his support and copies of useful material.

Mark Hudson, for his wonderful evocation of life in a former mining community – *Coming Back Brockens* – which won the A T & T Award in 1994, and for agreeing to write the Foreword.

Finally, to my wife Rosemary for her unfailing advice and encouragement.

Responsibilty for any errors in the use of all this advice and information is of course mine alone.

David Tonks
Stafford, November 2002.

Second edition

My thanks go especially to the Society's Graphic Designer, Graham Collett. His assistance, imagination and attention to detail has greatly improved the 'look' of the book.

I am grateful to the Society of Genealogists for their confidence in the need for a new edition. Interest in family history continues to grow and, as more and more material becomes available over the Internet I have tried to mention the websites that I have come across. It is impossible, however, to keep track of all such sites.

Thanks are again due to the archives, libraries and museums that responded to my queries about changes in their services since the first edition. One new feature is that almost all of them have online catalogues or guides. The three national coalmining museums have also extended their facilities for the family historian.

The Internet can never cover everything, and the collections held in all these institutions is often invaluable to a family historian with coalmining ancestors.

May we all continue to successfully 'mine' for information on our families.

David Tonks
Crook, April 2010.

Symbols

Throughout this book, the following symbols have been used:

	Book or article
	Exhibit
	Records
	Postal address
	Telephone number
	URL or web address
	Information
	CD ROM
	Archive or Record Office
	Library
	Museum

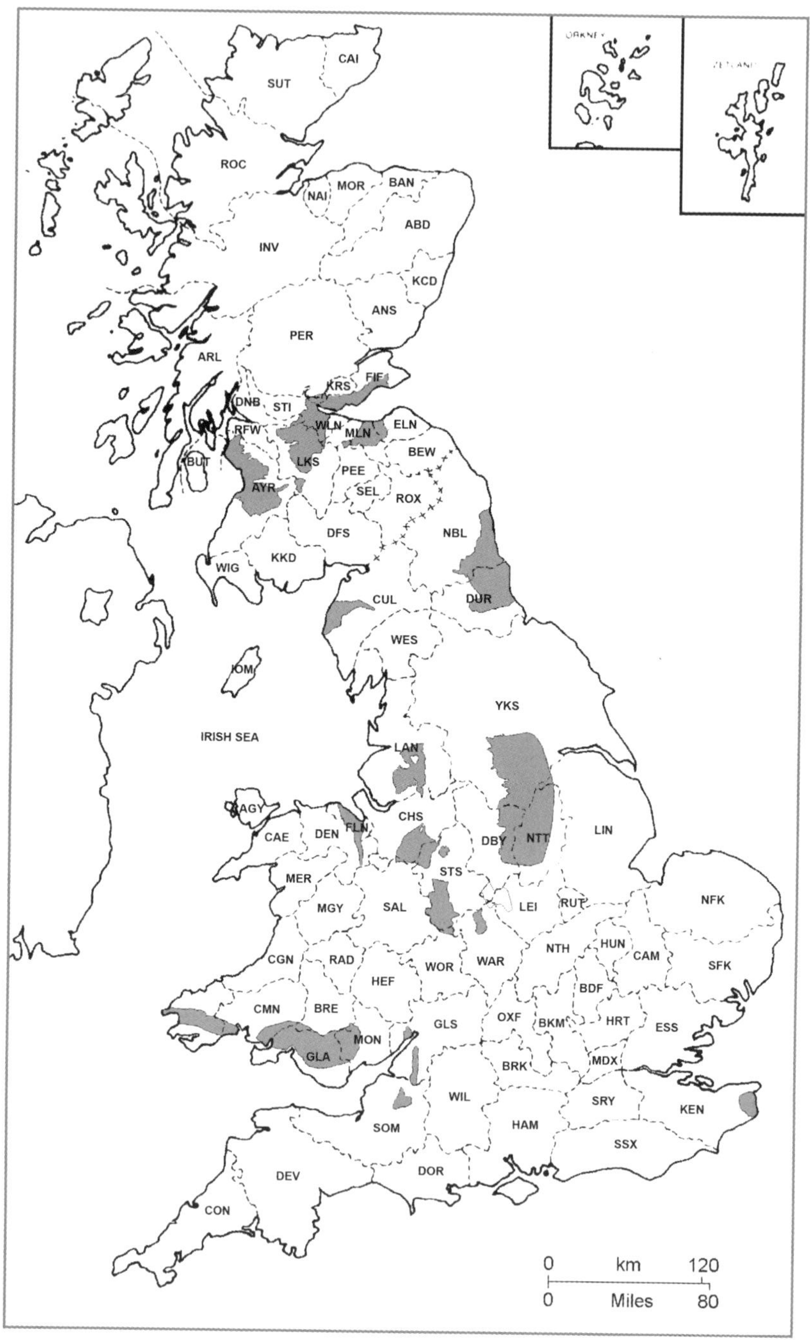

ORKNEY
SHETLAND
CAI
SUT
ROC
NAI
MOR
BAN
ABD
INV
KCD
ANS
PER
ARL
KRS
FIF
DNB
STI
WLN
MLN
ELN
RFW
LKS
BEW
BUT
PEE
AYR
SEL
ROX
DFS
NBL
WIG
KKD
CUL
DUR
WES
IOM
YKS
IRISH SEA
LAN
AGY
CHS
CAE
DEN
FLN
DBY
NTT
LIN
MER
STS
MGY
SAL
LEI
RUT
NFK
NTH
HUN
CAM
SFK
CGN
RAD
WOR
WAR
BDF
HEF
CMN
BRE
GLS
OXF
BKM
HRT
ESS
MON
GLA
BRK
MDX
WIL
SRY
KEN
SOM
HAM
SSX
DEV
DOR
CON
0   km   120
0   Miles   80

CHAPTER 1

Coalmining in Great Britain

Formation

To quote *Encyclopaedia Britannica*, coal is a 'solid, dark-coloured, carbon-rich material that occurs in stratified, sedimentary deposits. It is the most important of the primary fossil fuels and owes its origin to the partial decomposition and chemical conversion of immense masses of organic matter'[1] – such as trees with soft trunks, and hollow stems, or giant ferns. After the trees and plants died they formed swampy layers of peat-like material and this process was continually repeated. When the bottom of the swamp sank, rivers brought in sediment and, under pressure, formed bands or seams of coal submerged in seawater. Massive upheavals of the earth's crust brought some of these 'coal measures' to the surface as outcrops; others remained underground for thousands of years[2].

 www.secretshropshire.org.uk/Content/Learn/Mining/Geology.asp

Early history

Archaeologists have found evidence that coal was burned on Bronze Age funeral pyres in Wales; and that, when wood was in short supply, the Romans used it for metal working, heating, cooking and the drying of corn[3]. Traces have been found in Roman remains in the North East, especially near outcrops. It is also possible that the hot springs of Bath in the second century were fuelled by Somerset coal[4].

Coal is not mentioned in Domesday Book of 1086. London's first coal arrived by sea from Fife and Northumberland in 1228. In 1257 Queen

Eleanor was forced to leave Nottingham 'owing to the smoke of the sea-coals'[5] . By 1356 coal mining was well-established at Whickham in Co. Durham[6] and two hundred years later the North East coalfield was still the only one of more than local significance, output reaching about 40,000 tons per year[7].

A fifteenth century heraldic brass in Newland Church in Gloucestershire depicts a Forest of Dean free miner who wears a cap and carries a candlestick between his teeth. In his right hand is a small mattock, while a mine-hod of wood hangs at his back from a shoulder strap fastened to his belt; his leathern breeches are tied with thongs below the knee[8].

Plate I - Fifteenth-century Forest of Dean miner.

Sea Coal

At the beginning of the seventeenth century it was reported that:

> 'sea-cole and pitt-cole is become the general fewell of this Britaine Island, used in the houses of the nobilitie, cleargy and gentrie in London and in all other cityes and shires of the Kingdome as well as for dressing of meate, washing, brewing, dying as otherwise' [9].

A shortage of timber forced a change to this 'sea coal' as a domestic fuel, with annual output rising accordingly from 210,000 tons in 1560 to 2,985,000 tons by 1700. London was the main market and by 1720 over 60% of the output of the North East coalfield was being shipped there[10]. By the 1780s about 1.6 million tons of coal were transported by coaster. This had risen to over 13 million by the 1880s and 20 million by 1913[11].

Industrial Revolution

From about 1700 coal extraction became widespread in the West Midlands, Nottinghamshire, south west Lancashire, west Cumberland, South Wales, the Forest of Dean and Somerset. The small independent pit was the characteristic unit of production in these areas, with a workforce of about ten or twelve[12]. As the Industrial Revolution began to take effect, new machines such as the Newcomen engine for pumping, and steam-driven winding machinery, enabled the sinking of deeper and deeper pits. By 1791 Arthur Young was able to observe that 'all the activity and industry of this kingdom is fast concentrating where there are coalpits' [13]. It would be more accurate to say that it was the availability of coal plus other raw materials, which was the foundation of industrial development. On the Forth, for instance, the presence of coal and sea salt led to the establishment of a sulphuric acid works in 1749. In many cases coalfields were also ironfields, with 90% of all British pig-iron being smelted in the South Wales coalfields by 1806[14].

 Ashton, T S and Sykes, J *The coal industry of the eighteenth century*. 2nd. ed. (Manchester: Manchester University Press, 1964).

Transport

In the eighteenth century, coalmining interests in Lancashire did much to promote river navigations, turnpike roads and canals [15]. The development of mining and of railways proceeded hand in hand in the first quarter of the nineteenth century. Waggonways to transport coal were the forerunners of the railway and the original purpose of the pioneering Stockton and Darlington Railway, which opened on 27 September 1825, was to transport coal between the collieries of southwest Durham and their markets at Darlington and Stockton.

Nineteenth-Century development

Coalmining grew rapidly from about 1830 to 1870 to meet the increasing demands of industries such as brewing, cotton, iron and steel, milling, and pottery. By 1850 there were about 3,000 mines and output continued to rise – doubling between 1850 and 1880, and doubling again before the start of World War I. Some falling off in demand in the final quarter of the century, particularly in the iron and steel industries, was offset by an increase in exports. By 1870 over 11.5 million tons of coal were sold abroad, and by 1910 this had risen to 62 million. About a third of all coal output was exported by 1913. About three-quarters of this trade was concentrated in the South Wales and North East coalfields[16]. By 1900 there were over 1 million miners, and prospects for many more as developments continued in many of the traditional coalfields, and the newly discovered Kent field.

World War I

In 1913 a peak of output of 287 million tons was achieved and employment in coal mining was still high[17]. An advisory committee was appointed in February 1915 to ensure the availability of adequate coal stocks during the war. The coalowners were accruing huge profits, but miners' wages were no longer keeping up with the increased cost of living[18], and the miners of South Wales came out on strike. Unrest was beginning to spread throughout the country, and to ensure adequate wartime supplies, the industry was taken into state ownership. By the end of the war one of the benefits that resulted from this was the introduction of local pit committees, which provided an opportunity for the miner to participate, for the first time, in the management of the industry.

Decline

After the war many foreign markets were lost and the export price of coal fell by 50%. The coalowners' reaction to the general fall in demand was to reduce wage levels, and the miners responded by coming out on strike. However, the lack of support from other unions, on 'Black Friday', 15 April 1921, led to the capitulation of the miners in the summer of 1921[19].

Contraction in the industry accompanied by increased foreign competition and worsening industrial relations, culminated in the General Strike of May 1926 and the long coal stoppage (see Chapter 3). Around a third of a total workforce of twelve million went on strike, organised by the T.U.C. After nine days the strike was called off leaving the miners, once again, to fight the battle alone[20]. Finally, after a ten-month stoppage in the coalfields, the resistance of the miners collapsed.

The production of coal never again reached the levels of 1913. In 1933 only 207 million tons were produced, by half a million men – a reversal of two hundred years of steady expansion. After 1927 there was a steady closure of pits, but supply always seemed too great for demand until rearmament got underway in the late 1930s. Closures produced gains in productivity and new technology was adopted. In 1936, 55 per cent of British coal was cut mechanically compared to only 8 per cent in 1913. Investment was, however, always too little, and coal-mining areas were amongst the worst affected by structural unemployment[21].

 http://uk.encarta.msn.com/encnet/refpages/RefArticle.aspx?refid=781530891

World War II

After the years of depression, the industry was poorly placed to meet the demands of a wartime economy and by May 1941 a national coal shortage for the coming winter was forecast.[22]. The same month saw the application of the Essential Work Order in an effort to retain labour in the industry. This was a very unpopular measure in traditional coalmining areas. It was seen to bind workers 'more closely to the industry than had been known since the days of the annual bond in the North of England and of mining bondage in Scotland, more than a century before'[23]. Government control was deemed essential to manage the industry, but could not prevent a fall in production from 231.3 million tons in 1939 to 174.7 million tons in 1947. The reasons for this included lack of investment, shortage of materials, poor management, war weariness, and the loss of miners to the armed forces [24]. One attempt at reversing this trend was the Bevin Boys scheme, discussed in Chapter 7.

Nationalisation

1 January 1947 was 'Vesting Day' when the industry was nationalised at a cost of over £200 million[25]. The National Coal Board assumed responsibility for 980 mines and licensed a further 400 small mines in private hands[26]. The miners were jubilant and saw a brighter future in prospect, and the first Minister of Fuel and Power, 'Manny' Shinwell, anticipated a true partnership between himself, the Board, its managers and its mineworkers[27].

 www.num.org.uk/?p=history&c=num&h=9

Opencast mining

Although coal was obtained from 'open works' in the South Staffordshire coalfield from the 1840s, it was a hundred years later before opencast mining was officially established. Many artifacts relating to coalmining in former times have been discovered, such as a fifteenth century wooden shovel and an eighteenth century wooden sledge, which have been deposited in local museums; as well as examples of early bell pits and pillar and stall workings[28].

Local opposition to new sites is becoming increasingly bitter, with many concerns about health problems. In 2006/7 almost 1,700 men were employed in opencast or 'surface' mining, producing 8.4 million tonnes[29]. The Scottish Mining Museum has a Fact File on opencast mining.

 www.scottishminingmuseum.com

 www.ukcoal.com/sm-facts

Further decline

During the fifties and sixties levels of production were much improved after the introduction of mechanical aids. Unfortunately industrial and domestic consumers were beginning to turn to alternative fuels and the stockpiles of coal were mounting. The miners lodged a large pay claim but the Government supported the N.C.B, and in January 1972 the first national coal strike since 1926 began[30]. Mass picketing made coal stocks inaccessible, and it was this, rather than a shortage of coal itself, which closed many power stations. The Government declared a state of emergency, with powers to put industry on a three-day week as part of a campaign to conserve energy. However, it was soon forced to capitulate and concede big pay rises. Miners' earnings in 1972 jumped up by 16 per cent, more than double the rate of inflation.

Public sympathy was with the miners, but the industry was to pay the price. The age of cheap energy had dawned, and pit closures occurred at the rate of about one a week so that by 1974 only 246 pits remained and employed only 246,000 men[31]. A sharp increase in the price of oil led to a brief recovery and the 1974 *Plan for Coal* suggested an increase of 42 million tons per year by the mid-eighties.

Rundown

By 1983 it was obvious that such targets were over optimistic as industrial customers had continued to switch to other forms of energy. The NCB was losing £143 million per year, and further rationalisation was called for. This led to the bitter dispute between the miners and the government of 1984-5 in which about 38 million working days were lost (see Chapter 3). Compromise between the two sides appeared impossible, both being convinced of the overwhelming strength of their case. It was a period of suffering and privation for mining communities, where many miners' wives played an active and political role in supporting the strike. Eventually and inevitably there was a return to work, and to pit closures and flexible working.

Between 1984 and 1994 employment in deep mining fell from 181,000 to 10,000, as the number of collieries fell from 170 to 16 [32]. The NCB became British Coal in 1986. By the 1980s only 35% of the energy needs of the country were being met by coal; in the 1940s this figure was over 90% [33]. In October 1992, British Coal announced a programme of 31 colliery closures, and in 1995 employment in the industry was no more than 2% of those employed in 1947 [34].

Privatisation

In December 1994 the mining industry was privatised; RJB Mining paid the British taxpayer £814 million for the English coalfields and became the UK's biggest coal producer. In 2000 there were seventeen deep mines in production in the UK, producing about 25 million tonnes of coal a year, and employing some 10-15,000 people. At the end of 2007 British Coal (the company's new name) operated four deep mines, located in Central and Northern England, with an output of around seven million tonnes and employed 2,400 people[35].

Into the new Millennium

Although a number of government initiatives were implemented, designed to provide new jobs and new hope for mining communities, later research found that coalfield communities remained blighted by widespread unemployment, long-term sickness and poverty[36].

The mining industry continued to decline, although fears about the availability of other sources of energy, such as oil and gas, lead to calls for possible expansion, in deep- as well as opencast mining. Research continues into making the burning of coal in power stations a more 'greener' option. For the year ended 31 March 2007 it was reported by The Coal Authority that coal output was a mere 8.2 million tonnes from nineteen deep mines employing 3,600 workers.

Online resources

 'The Story of Coal' can be viewed or downloaded from the British Coal website: www.ukcoal.com/story-of-coal.

 A 'learning module' from the Science Museum on the coal industry: www.makingthemodernworld.org.uk/learning_modules/history/02.TU.02/ ?section=8.

CHAPTER 2
Why is he different?

The first stages in tracing a family history which includes a coalminer are no different to the techniques employed for any other family. All the standard sources should be consulted, perhaps while following the advice of one of the many excellent textbooks or magazine articles which are now available. It is an almost impossible task to pick out the most suitable from this vast array, but two of the most comprehensive (and which also contain sections on coalmining) are:

 Heber, M *Ancestral trails: The complete guide to British genealogy and family history*. 2nd ed. (London: Genealogical Publishing Company, 2006).

 http://www.ancestral-trails.com/

 Barratt, N *"Who Do You Think You Are?" encyclopedia of genealogy: the definitive reference guide to tracing your family history*. (London: HarperCollins, 2008).

Nick Barratt has also contributed 'Family History: getting started' to the BBC Family History website. The 'Who Do You Think You Are' television series also has a website.

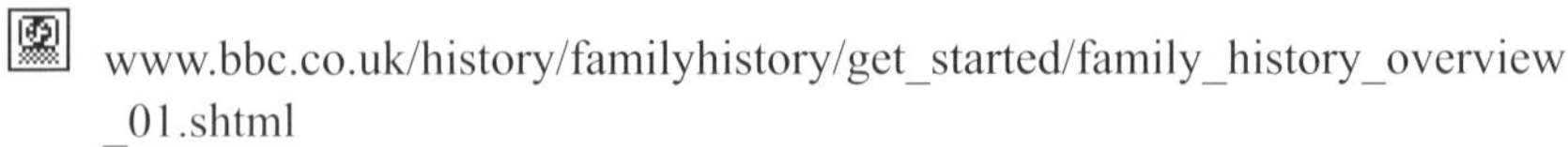 www.bbc.co.uk/history/familyhistory/get_started/family_history_overview
_01.shtml

www.bbc.co.uk/whodoyouthinkyouare/

So one must consult elderly relatives; visit record offices and libraries; search the Internet and glean as much as possible from birth, marriage and death records; church records; the various formats and editions of the IGI, between 1841 and 1901; the Census Enumerators' Books and, for 1911, the Household Schedules and Enumerators' Summary Books.

familysearch.org

www.nationalarchives.gov.uk/census/

http://freebmd.rootsweb.com/cgi/search.pl

One should also consider joining a family history society, either locally, or one (or more) covering the mining areas where your ancestor lived and worked,

www.ffhs.org.uk/members2/alpha.php

and the vast resources available on the GENUKI website should not be overlooked:

www.genuki.org.uk

There are a small number of published books on family history for those with coalmining ancestors. As well as the first edition of the present book, these include:

The Guardian guide to family history (London: The Guardian, 2007). Includes a case history relating to a coalmining family that is also available on the Internet at:

http://lifeandhealth.guardian.co.uk/guides/familyhistory/story/
0,,2053946,00.html

Henesey, A *Routes to your roots: mapping your mining past* (Overton: National Coal Mining Museum for England, 2004).

Reeks, L S *Scottish coalmining ancestors*. (Baltimore: Gateway Press, 1986).

In addition, it would seem that about once a year one or other of the family history magazines includes an article specifically on coalmining and coalmining ancestors. Such articles often provide further guidance for the beginner.

Finally, the Durham Mining Museum website includes some useful advice for the family historian.

 www.dmm.org.uk/famhist.htm

Once this primary stage is complete, or when faced with seemingly insurmountable hurdles whilst attempting to do so, that these questions may be asked:

+ 'Why is he different?' or,

+ 'what were his *distinctive* traditions and characteristics?'[1], or, more importantly,

+ 'how can I capitalise on these to make my family tree as detailed and interesting as possible'?

It is perhaps easier to list those sources in what might be called the secondary phase which do *not* apply, than those which do. There are *unlikely* to be mentions of a coalmining ancestor in, for instance, trade directories (unless he was an official or coal owner); or poll books before universal suffrage (he rarely owned property and therefore had no right to vote); or in the published register of a public school or university (although some examples do exist).

However the coalminer – his work, lifestyle and family circumstances – was sufficiently 'different' to open up several new areas of inquiry for the family historian. These may be grouped and summarised as follows:

+ *Home*
     Housing
     Lodgers
     Migration
+ *Family*
     Age at death
     Church records
     Family size
     In-breeding

- *Work*
    The Bond
    Butties and Truck
    Disaster victims
    Serfdom
- *Government*
    Government Reports

Housing

Most historians of the industry warn against attempting to quantify the income of a miner, certainly until nationalisation began to introduce standardisation of wage rates and working conditions. Before that there was little certainty about the contents of the wage packet from pay-day to pay-day. However, an important component of the miner's household economy was his receipt of a 'free' house, or at least a rent allowance. Although this concession did not operate in all coalfields, nevertheless from the early days it applied to a very great extent in Ayrshire, the Lothians, and Northumberland and Durham. In 1800 it was reported that:

> 'in almost every colliery on the Tyne pitmen are provided with houses by their employers, for which, and about 8 or 10 cart-loads of coals annually, they pay 3d. a week'[2].

As the industry was developing, often in virgin countryside with few, if any, amenities, it was necessary for the coalowner to provide housing for his workforce. These houses he could either build himself, or employ a contractor to build them on his behalf. They were then let on minimal rents to the miner and his family as part of his 'contract of employment', and there are obvious similarities here to the agricultural 'tied cottage'. Although the miner was no doubt grateful for a roof over his head, the colliery house was to become one of the causes of controversy and discontent which would affect the industry in the coming years. Evictions during disputes are known to have occurred in Scotland, the North East and Yorkshire, even into the twentieth century. In some coalfields the employer provided houses only as an attraction when the pit first opened, and as a means of retaining key workers [3].

In 1915 Jevons reported that development was being 'seriously hindered' in the South Wales valleys by the lack of houses, miners having to travel 15-20 miles by rail or to walk 3-4 miles over the hills in all weathers. He described a 'house famine' in the rapidly developing fields of South Wales and South Yorkshire in this pre-First World War period [4].

The provision of housing for miners is described in more detail in Chapter 4.

The tradition of providing housing for retired miners can be traced as far back as 1701, when the keelmen of the Tyne, whose boats transported coal downriver, opened a residential home for retired and disabled members and widows[5]. However it was the miners of Boldon, south of the Tyne, who in the 1890s first proposed the provision of accommodation for their retired colleagues. This led to the formation in 1898 of the Durham Aged Mineworkers' Homes Association which, at the time of its centenary, owned some 1,250 houses for retired miners.

 Atkinson, G L *The miners' heritage: a history of the Durham Aged Mineworkers' Homes Association 1898-1997*. (Durham: County Durham Books, 1997).

 www.durhamhomes.org.uk/index.htm

Lodgers

If the problem of housing was severe for the miner and his family, especially in the newly-developing coalfields, then the accommodation of lodgers was even more difficult. Young single men have always been attracted to the comparatively high paid jobs in mining, especially in the first phase of development of a new field or colliery. In many cases the only way to house these men was as lodgers in the households of fellow-miners. In 1842 there were Lanarkshire families in two-roomed cottages who were taking as many as fourteen single men as lodgers, but it was a long-standing and national problem. By 1911 almost one tenth of all houses in the coalfields of the North East and South Wales were being sub-let [6]. This was inconvenient for householders and lodgers alike, especially if the workers were on different shifts. However the income from a lodger could help out with the household expenses of a large family.

Many coalmining ancestors, therefore, are likely to be found as lodgers in the household of another miner's family, and this could be one explanation of 'missing' relatives. To find them use one of the many census indexes (printed, on-line or, on CD-ROM). These tools are also invaluable if your ancestors moved about the country.

Migration

Most studies of migration - 'the permanent or semi-permanent movement of individuals, families or groups from one place to another'[7] – emphasise the importance of economic considerations. Most individuals and families move from country to town for reasons related to employment.

Early coalmining was carried out by villagers working on land which happened to have seams at or just below the surface. As technology and the market for coal advanced hand in hand, collieries became larger and more sophisticated, pits became deeper, and the number and range of mining jobs increased many times over. This work could not be undertaken just by the local inhabitants and so more and more men were attracted to work in the industry. The skills employed in other branches of mining, such as for tin or lead, were often transferable to newly expanding coalmines in other parts of the country. During the nineteenth century this sequence of events was repeated over and over again, as new coalfields were developed, new pits sunk, new communities established. It has even been estimated that up to a tenth of all miners moved house every year [8].

Plate II - Moving house.

Sometimes families would move only a few miles if a new colliery was being developed close by. On other occasions the distance could be tens or even hundreds of miles, especially as the older coalfields began to run down. New pits were opening elsewhere in the country such as those in Fife, Kent, Yorkshire, or South Wales where, at the time, conditions were superior and higher wages were paid [9]. South Wales was the biggest draw with nearly 100,000 workers coming to that coalfield between 1901 and 1910 from Cornwall, Gloucestershire, Somerset, the West Midlands, and the rest of Wales itself [10]. So a coalmining ancestor could well have lived in different parts of the country before settling-down for the last time. This could well explain why 'home' was always some distance away.

One way of following these family moves is to examine the birthplaces of the children in mining families. The entries in the Census Enumerators' Books will often show that each child was born in a different place, perhaps even a different county, as can be seen in the example on page 73.

A famous Durham miner provides an even more interesting example. The great miners' leader, Peter Lee, experienced migration at first hand, in childhood and as a working miner:

'Memories of my early days are of moving from village to village, viewing our English lanes from the top of a wagon-load of furniture ... This moving from place to place was a very common experience in our lives.'

As his biographer, Jack Lawson, described:

'Between 1879 and 1886 Peter Lee worked at fifteen collieries. Here is the tale of it. He worked at Littletown twice, The Boyne, Pittington three times, Sleekburn, Houghall, Elemore, Haswell three times. Two pits in Oldham, twice at Wingate, Trimdon Grange, Trimdon, Littleburn and at two pits in Dearham, Cumberland' [11].

And another miner writes:

'Many thousands of families would change districts annually, and in April and May the whole mining districts were alive from side to centre with loaded vehicles from 'big waggons' to 'cuddy carts" [12].

This frequent up rooting of miners and their families reduced as the industry began to stabilise, although new mines, as and when they opened, still attracted miners from other parts of the country. In the final quarter of the twentieth century, when the industry began to decline and pits close, often the only course open to a man who wanted to stay in the industry was to move to an area where mining was still being carried out.

 Camp, A J *My ancestors moved in England or Wales: how can I trace where they came from?* 2nd. ed. (London: Society of Genealogists, 1994).

Age at Death

That mining was a dangerous occupation there is no doubt, and this topic is explored further in Chapter 3. In all the major British coalfields, deaths from accidents were around three times as frequent as in the population generally. In the period 1860-1897 mining accidents left some 13,000 widows and over 30,000 children [13]. Only fishermen and workers in stone and slate quarries suffered from similar losses.

However, after the results of the 1881 Census were analysed, it was reported by the General Register Office that:

> 'The death-rates of miners are surprisingly low. In spite of the terrible liability to accident and their constant exposure to an atmosphere vitiated by coal dust, by foul air and by excessively high temperatures, the comparative mortality figures of these labourers are considerably below that of all males'[14].

Although it can be argued that in general terms the overall health of coalminers improved over time, there were a number of occupational diseases which resulted from their work – such as bronchitis, rheumatism, pneumoconiosis and silicosis, as well as minor complaints such as 'pit knee ' and 'beat elbow'[15]. That these conditions were common in the mid-nineteenth century, and that former miners were still affected by them in the twenty-first century, is a pointer to the unhealthy conditions which all miners had to endure, and which often led to their early demise.

Church Records

There was no standard form for any parish register until Hardwick's Marriage Act of 1753 and Rose's Act of 1812. The latter came into force on 1 January 1813 and standardised the forms of entry for registers with which we are now familiar. However, there are examples of registers which provide more information than those post-1813. In 1798 Shute Barrington, Bishop of Durham, instructed his clergy to provide more detailed registers in their parishes. As his diocese covered the whole of the Northumberland and Durham coalfield, those with coalmining ancestors from the North East who were baptised or buried in the period 1798-1812 may benefit from his innovation.

The Registers of Baptisms may include:

- name and place of birth of the child

- date of birth

- date of baptism

- number in the family (1st Daughter; 2nd Son, etc.)

- name, occupation and parish of origin of the father

- maiden name and parish of origin of the mother

- name and parish of origin of the mother's father.

The Registers of Burials may include:

- name and place of death

- occupation of the deceased

- date of death

- date of burial

- age at death

- name, parish of origin and occupation of spouse

- for a child – parents' names, occupation of father and maiden name of mother.

So it is possible, after locating just one baptism or burial, to ascertain a great deal of information about a particular family, and it is regrettable that these details cannot be obtained so easily from the registers after 1813. Examples of these registers may be seen at the record offices of Durham, Northumberland and Tyne and Wear (see pp. 92-93).

It should of course be remembered that although many coalminers at this date and in this area, and elsewhere, may not have been active churchgoers, they still attended the Anglican parish church for the rites of baptism, marriage and burial. They may have been staunch Non-conformists or Roman Catholics, but nevertheless their names will be recorded in the parish registers, probably at least until civil registration commenced in 1837 and often for many years afterwards (see Chapter 5).

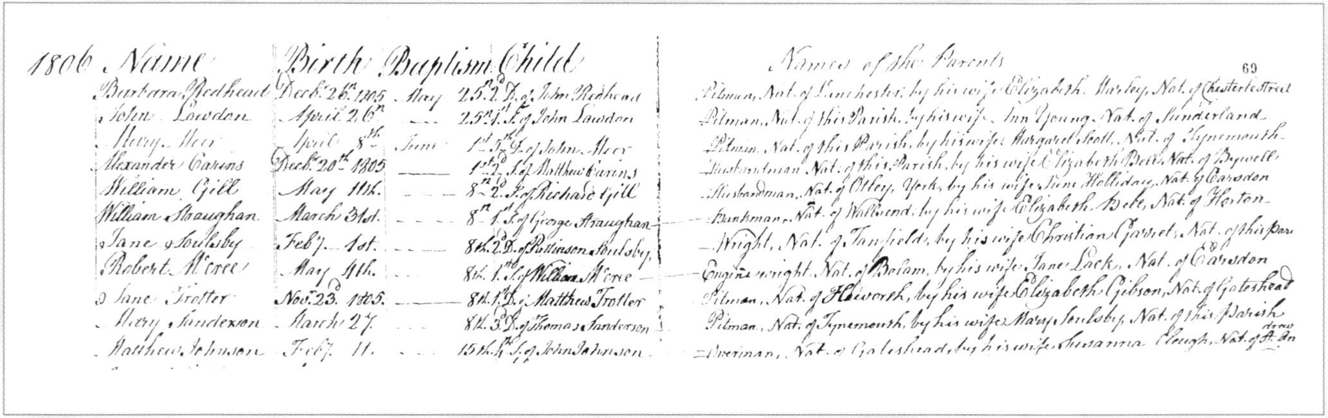

Plate III - 'Barrington' Baptismal Register, Longbenton 1805.
(Reproduced by permission of Rev. P.S. Ramsden, Vicar of Longbenton).

 Leary, W *My ancestors were Methodists*. 3rd. ed. (London: Society of Genealogists, 2005).

Steel, D J *National index of parish registers, vol. 1, Sources of births, marriages and deaths before 1837 (1)*. (London: Society of Genealogists, 1968). pp. 44-45.

Family size

The theory that coalminers had large families, and the reasons why this occurred, continues to be discussed and considered by social as well as family historians [16]. Stereotypically, they were careless, feckless, unthinking of the future, or trying to ensure the family income in the event of an accident to the main wage-earner. However, it now seems likely that other occupational groups also had large families. The 1911 Census revealed that, although the average mining family had 4.23 children, compared with the national average of 3.53, others such as general and agricultural labourers had even more children[17]. However, even a cursory examination of the Census Returns for any mining community will reveal families of five, six, seven or more children (see page 73). Accommodating them, plus a lodger or two, in a tiny pit cottage can only have been accomplished with considerable ingenuity. Miners continued to have large families: R.H. Shephard of Denaby Main, who married in 1911, was a father of ten[18].

In the early days there was no doubt of the value of mining sons. To his description of family life in 1855, Leifchild adds a footnote:

'Families of *boys* are, amongst pit-people, valuable property, on account of their earnings in the pits. A widow with a family of boys is considered *a catch*. I was told that such a widow was accosted by a suitor even at her husband's grave. Her reply was, 'you are *too late:* I am engaged. I accepted B----- before starting for the funeral'[19].

In-breeding

Leifchild described miners as 'a distinct race of beings':

'His stature is diminutive, his figure disproportionate and misshapen; his legs being much bowed; his chest protruding (the *thoracic* region being unequally developed.) His arms are long, and oddly suspended. His countenance is not less striking than his figure; his cheeks being generally hollow, his brow overhanging, his cheekbones high, his forehead low and retreating; nor is his appearance healthful; his habit is tainted with scrofula'[20], and this he ascribes not only to his physical labours, but also to his habit of in-breeding, which has transmitted these defects 'through a long series of generations'. As a result of the isolated communities in which they lived and, in Scotland, their serfdom, there was a tendency for miners' sons and daughters to marry. However, whether they were already related is less clear; and it is also debatable if Leifchild's theories would be acceptable to modern genetic theorists.

Nevertheless the bowed-legged, hollow-cheeked pitman could be seen in many a mining community one hundred years after Leifchild described them, and the probability of an ancestor's mother's father, as well as father's father, being a miner must be significant.

The Bond

In the pre-Industrial Revolution economy of England and Wales miners, like many other workers, were contracted for a year at a time. This security of employment on an annual basis had advantages for employer and employee, not least being the immunity from wartime recruitment. Payment was usually made on hiring day- from 2*s*. up to 10*s*. 6*d*. - and, as in Lancashire in 1676, the miner undertook to:

> 'well and truly serve for a year in the coal mines at Hulton or Denton according to the customs and orders there or to forfeit £40 sterling'[21].

However by the beginning of the nineteenth century, the system of binding – the Bond – was generally limited to the Northumberland and Durham coalfield, where drinks were provided and hewers could receive up to five or even ten guineas on binding day. At one colliery in 1800 men were 'kept ... well supplied with Ale so long as they would drink it'[22]. Binding time varied over the years between October, January and April. The terms of the Bond were also subject to variation, as the miners began to come together in early trade unions and struggle each year for better working conditions and higher wages, all of which were regulated by the Bond. There were many disputes on this issue, in 1826 over a standard Bond, in 1831 and 1832, and finally and crucially in 1844 when the system was virtually brought to an end by a strike of some twenty weeks. In 1869 it was abolished in the Courts due to the efforts of 'the Pitman's Attorney', W.P. Roberts.

Some bonds are still extant and contain the signature (or mark) of the miners agreeing to be bound. Examples can be inspected in, for instance, the record offices of the North East and some are reproduced on the website of the Durham Mining Museum.

 www.dmm.org.uk/archives/index.htm

Butties and Truck

In some coalfields these two often inter-related phenomena, especially after around 1850, were common, and both had an effect on the wages which the miner received. In the butty system the underground employees (other than management) were employed by a sub-contractor – known as the charter-master or more colloquially the butty – and not the mine-owner. This system operated particularly in the Black Country but also in the Yorkshire, Derbyshire, Nottinghamshire, Lancashire, Shropshire, and North and South Wales coalfields. Its main advantage to the coalowner was that he was required to provide less working capital, some capital being provided by the butty. But both owners and miners came to see the butty as a slave-driver who benefited from their labours. In 1879 one owner wrote:

> 'The men who do the hardest work can, under this vexatious system, scarcely get enough to exist upon, while the butty is feasting upon his ill-gotten-gains'[23].

By the later years of the nineteenth century, as mines became deeper and collieries more sophisticated – and therefore expensive – butties no longer had the necessary capital, and the owners' investment was too substantial, not to be fully in control of all aspects of production. However it did not fully come to an end until the beginning of World War I, and even many years after, in some coalfields[24].

Truck – the system of payment in kind, rather than cash, was not unique to coalmining areas, but legislation outlawing it in such areas was rather slow in coming, and when it did eventually arrive was not wholeheartedly supported by magistrates. In 1822 it was reported that:

> 'By this practice Coal and Iron masters compel their workmen to accept of two-thirds of their wages in goods, such as Sugar, Soap, Candles, Meat, Bacon, Flour, etc., instead of money, at an unreasonably large profitt'[25].

Truck was one of the many grievances that brought miners out on strike in the early nineteenth century. Originally, however, the provision of a colliery shop (or Tommy Shop) was considered as essential as that of colliery houses, and does not appear to have been imposed or exploited until the 1820s. Sometimes, especially in Scotland and South Wales, the high prices charged in the shop were offset by high wages. Although it was officially abolished in 1831, it continued for at least another fifty years in some coalfields, being finally outlawed by the Truck Amendment Act of 1887.

The shops were operated by the owners themselves, or by the butty, who might also keep a public house and encourage the miner to spend his wages there, as was admitted by a former Black Country butty in 1843:

> Most of the butties keep public-houses, and if you don't pay your footing you are not employed; you must go without victuals, or else pay for ale[26].

Disaster victims

The most common of all specific records relating to coalminers is probably the sad occurrence of the recording of a death due to an explosion or accident. The names of victims are recorded in various national and local sources, see Chapter 10.

 Winstanley, I *Mining deaths in Great Britain*. 7 vols. (Wigan: Picks Publishing, 1995).

 www.cmhrc.co.uk/site/disasters/ *A database of over 164,000 records containing the details of coalmining accidents and deaths in the UK.*

The appropriate local record office may also include such records, as well as of the families of victims who were the beneficiaries of any Relief Fund which may have been set up at the time. Examples include the Sheepbridge Coal and Iron Co. accident record books at Derbyshire Record Office and Lea Green Colliery miners' relief society minutes, 1933-64 at Lancashire Local Studies Library and Record Office. Some records of this nature have been published.

Serfdom

Contracts between miner and owner, similar to, but more stringent than bonds, existed in parts of the Scottish coalfield. From 1606 coal and salt workers were forbidden by law to leave one master for another on pain of fines or corporal punishment. Miners were specifically excluded from the Habeas Corpus Act of 1701, and further Acts and decisions of the Courts legalised what has been described as 'a degradation without parallel in the history of labour in Scotland' [27], and created an era of 'serfdom' for Scottish miners from 1606 to 1799. 'Arles' - gifts of cash or kind – were a token of engagement, and were even given by the master as a baptismal gift to the child of a miner, ensuring that he, too, was bound into serfdom. The main aim of the system was to limit the mobility of the miner, and it was first introduced when there was a shortage of labour in the coal and salt industries. In fact owners seem in fact to be able to attract (or poach) miners from other collieries, with miners moving almost as freely as in England and Wales.

Opposed by many coalowners, an Emancipation Act was passed in 1775, although it did not grant automatic freedom, which had to be sought in a Sheriff Court. Another Act was introduced in 1799 as 'many colliers still continue in a state of bondage' [28].

Government Reports

Because of the importance of the coal industry, it is hardly surprising that the number of published reports from government-appointed committees or commissions is legion. Such reports often led to new legislation, which itself spawned even more reports. One of the most useful to the family historian is the first report of the Children's Employment Commission – on Mines – which was published on 7 May 1842. The report and its appendices caused an enormous stir on publication. It was one of the first government reports to include illustrations, and these now famous engravings of underground scenes, depicting the terrible hardships endured by boys, girls and women, added to the sense of outrage.

If it can be established that an ancestor was employed in coal mining at this period, it is well worth checking these volumes; you may find the most vivid description, sometimes in their own words:

No. 108.—William Simpson. Aged 6 years.

'Has just been down a week. Keeps a door. (George Kendall works besides him, and states that he is very frightened when his low (candle) goes out, and cries out. Is 12 hours down the pit, and works in the night shift now). Cannot read or write. Was at school a short time. Goes in about a mile by' [29].

The three volumes have been published in facsimile by the Irish University Press and may be available in the larger public reference and university libraries. Local studies libraries and record offices may have copies of the reports on their local areas, and they have been scanned by Ian Winstanley and made available at his Coal Mining History Resource Centre website.

 www.cmhrc.co.uk/site/literature/royalcommissionreports/

The 'Women and the Pits' website includes excerpts from several areas and many local mining and historical websites have also included the reports appropriate to their area.

 http://freepages.genealogy.rootsweb.ancestry.com/~stenhouse/
coal/pbl/coalmain.htm

 www.scottishmining.co.uk/16.html

One of the outcomes of this report was the creation of an Inspectorate of Mines, whose reports were to be very influential in matters such as health and safety, education and training and many other reforms. Since 1842 reports and legislation have continued to pour from the government presses, and the former may well include verbatim evidence from a coalmining ancestor, or at the very least, provide useful background reading on the industry. Details of some reports may be found on the BOPCRIS website, which aims to provide access to British official publications, some with full text, over the web. If you are a member of a library that subscribes to *House of Commons Parliamentary Papers* then you will be able to search for and access over 5 million pages of these publications.

 http://parlipapers.chadwyck.co.uk/marketing/about.jsp

 www.bopcris.ac.uk/bopcris/digbib/home

Many books on coalmining quote from Victorian government reports, which provide a vivid picture of conditions at the time.

 Children's Employment Commission *First Report of the Commissioners: Mines*; *Appendix to First Report of Commissioners: Mines Part I Reports and Evidence from Sub-Commissioners*; *Appendix to First Report of Commissioners: Mines Part II Reports and Evidence from Sub-Commissioners*. 3 vols. (Shannon: Irish University Press, 1968), originally published London: H.M.S.O., 1842.

 Pike, E R *Human documents of the Industrial Revolution*. (London: Allen & Unwin, 1966).

Pollard, M *The hardest work under heaven: the life and death of the British miner*. (London: Hutchinson, 1984).
A persuasively argued and vividly written account of mining life, with much detail gathered from the Children's Employment Commission Report.

CHAPTER 3
Work

The Job

The earliest methods of coalmining involved no mining at all. Coal was just picked up from the surface, or from shallow ditches. Then drifts were dug into the hillside, following the seam. Later still came the bell pit, like a well shaft sunk down to a narrow seam. The coal at the foot of the shaft was removed, and then work extended until there was danger of the sides collapsing. The pit was then abandoned, and work began again. However it was not long before the practice developed of heading out into the seam for a short distance on each side of the shaft [1].

Naturally, some method of supporting the roof became necessary and this led to the practice of leaving pillars of coal in place as roof supports. Eventually this method of underground working became known as 'pillar and stall', 'bord and pillar' (in the North East), or 'stoop and room' (in Scotland), where two sets of roadways (stalls) are driven at right angles to each other, leaving coal in place (pillars) to support the roof. These remained as work continued outwards, but as the men began to work back towards the pit bottom they were removed and the roof allowed to fall (unless there were buildings on the surface above). This ensured that the

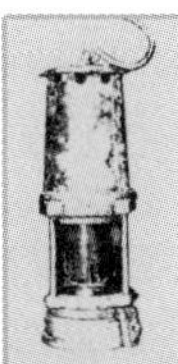

valuable coal was not wasted, and was known as working 'in the broken' [2], or, 'coming back brockens' [3].

The alternative method - 'longwall' - was where a number of men worked together along a face, and removed the whole of the coal seam in one operation. No pillars of coal were left in position and, to support the roof, their place was taken by wooden props [4]. There were a number of variations on these two basic methods of working, such as 'bank work' in Yorkshire and the East Midlands. 'Longwall' was to become the most popular method of working, except in North East England.

In the nineteenth century it was possible to move up the various grades of pit work, starting as a trapper boy and ending up as a hewer. This was certainly the case for George Parkinson in the 1830s:

'I passed from childhood to manhood through the ordinary curriculum of the northern pitboy's lot. I graduated successively from the starting point of a doorkeeper in the mine at nine years of age, through all the stages of a miner's toil and its dangers, till at twenty-one years of age I took my degree as a coal-hewer, this being the highest unofficial position attainable at the cost of the hardest form of mining-labour known'[5].

Throughout the country, in 1841, there were some 5,000 children between the ages of five and ten working underground, most of them probably trappers[6].

At this time their duties were as follows:

'He sits himself in a little hole, about the size of a common fireplace, and with the string in his hand: and all his work is to pull that string when he has to open the door, and when man or boy has passed through, then to allow the door to shut of itself. Here it is his duty to sit, and be attentive, and pull his string promptly as anyone approaches. He may not stir above a dozen steps with safety from his charge, lest he should be found neglecting his duty, and suffer for the same'[7].

At about fourteen a boy became a 'pony driver', leading a pony as it pulled a 'set' of tubs underground. At sixteen, as he became stronger, he moved on to 'putting', supplying the hewers with empty tubs, and moving full ones. He moved on again, to become a 'hewing putter', and the transition was complete at about the age of twenty-one, when he became a hewer, whose job it was to produce good, round, merchantable coal. The hewer could then ascend the ladder of promotion by study and examination, and become a deputy, an overman, or even an under-manager or manager.

Plate IV - Hewer working in a naked light pit in North West Durham, c1900.

However, for underground workers, the job of the hewer was the undoubted pinnacle. The 'big hewer' had a fearsome reputation as the hardest and most productive worker in the pit, whose skills were proven in hewing matches, and whose prodigious feats were commemorated in many a story or song. Here is a description of the hewer, taken from a *Glossary of terms used in the coal trade of Northumberland and Durham*, whose third edition was published in 1888.

> HEWER – A man who works coals. His age ranges from 21 to 70. His usual wages (1849) are from 3s.9d to 4s.3d per day of 8 hours working, and his average employment 4 or 5 days in the week. He also has, as part of his wages, a house containing two or three rooms, according to the number in his family, and a garden, of which the average size may be 6 or 8 perches; also a fother of small coals each fortnight, for the leading of which he pays sixpence.

Each colliery had a 'management team', consisting of a manager (or Viewer), and deputy- and under-managers. Below them, the officials of the colliery consisted of the shift supervisors, or overmen, and their 'deputies'. who were responsible for safety. All of these men would spend a greater or a lesser amount of time underground. But about a fifth of all colliery workers worked above ground in tasks as varied as winding-engineman, screensman, blacksmith, and other tradesmen. They all contributed to the output of the pit, but never received the attention, or the admiration, which the hewer attracted.

Plate V - Colliery officials, 1904.

Before the start of World War I the majority of the coal in British pits was got by the primitive, back-breaking efforts of the hewer. Work at the coal face had changed little in two hundred years, the only 'machines' in use being the handtools of the hewer, the brawn of the putter and the strength of the pit-pony, although coal-cutters, powered by compressed air, began to be introduced from the 1870s, and by 1910 improved machines powered by electricity were in evidence. By the 1930s the chain coal-cutter was the most widely used in all coalfields. The next stage in coal face mechanisation was the face conveyor. There were various types, including an endless conveyor belt of rubber or other less flammable material. A further development was the introduction of the first effective longwall power-loading machine, described as a 'harvesting machine' for coal. By 1970, over 90% of coal was won by power-loader machines [8].

In 1841 miner William Wardle of Eastwood described his job as 'the hardest work under heaven'[9], and the same description could readily be applied to miners before and since, right up to the present time. The dangers inherent in his employment are explored in the next section.

Danger

'Thousands upon thousands of you, however, are totally unacquainted with the real situation of this most useful, yet most abused class of men. You have no real idea of the dangers that every moment stand exposed to, while procuring the fuel which comforts, cheers and enlivens you around your social hearths'[10].

'The dangers which attend the coal-miners' work are of several different kinds, and unless the most strenuous and constant precautions to prevent accidents are taken both by the management of the mines and by individual miners, the hazard to life and limb is greater than in any other occupation'[11].

Although many years separate these two quotations, they seem to indicate that, as far as mines safety is concerned, little progress was actually made between 1831 and 1915. Of course this is untrue. Mines have been inspected since 1851, there were six inspectors by 1852, about 30 by 1900, 90 after the Coal Mines Act of 1911, and more than 160 in the 1950s [12]. Safety legislation was constantly amended and updated, yet the industry was still a dangerous one in which to work. The major disasters are considered separately below, but injury, mutilation and death were a common – almost everyday – occurrence in mining communities, although in fact there was an improvement decade upon decade, and during the many years when coal production and manpower levels were rising, the death rate was falling[13]. Nevertheless, over 100,000 miners have lost their lives since 1851.

It was these dangers which set the miner apart from workers in other industries, although it was only the accidents in which there was great loss of life which came to the attention of the general public. The miner was also prone to malignant diseases peculiar to his work, such as pneumoconiosis – a deadly disease of the lungs caused by the constant inhalation of dust; nystagmus – an extremely painful eye disease caused through working in a feeble light; bronchitis[14]; and 'beat knee' - caused by crawling and kneeling[15].

 Jevons, H S *The British coal trade*. (Newton Abbott: David & Charles, 1969), first published 1915. 'The Collier's Daily Toil'. pp. 608-617.

Disasters

Some 70,700 miners died or suffered injury in the mines of Great Britain from 1850 to 1908, according to the appendices of the Mines Inspectors Reports 1850-1914 [16]. It was only from the 1930s that there were fewer than 1,000 fatalities per year.

Given the horrific loss of human life in the industry, it is surprising that the first rescue station was not opened until 1902, and that 'Central Rescue Stations', within ten miles of every mine, were not made compulsory until the Coal Mines Regulation Act, 1911 [17]. At first, an important part of the equipment was a canary. These birds are highly sensitive to carbon monoxide, and were used to test for the presence of this deadly gas, before the introduction of electronic detectors [18]. There are many examples of the members of rescue brigades, and of colliery rescue teams, losing their own lives in the course of seeking to save others.

Although mining disasters were events of great terror to the victims and great sadness to the bereaved, they did nevertheless give rise to a number of different examples of contemporary evidence, which might well serve to enhance the history of a mining family and the community of which it was part. Some examples of the different records which are available are provided and described in Chapter 10.

The discoveries made during the investigations into the cause of many disasters often led to improvements in safety arrangements, providing a lasting memorial to those who had lost their lives.

Anderson, M *Durham mining disasters c. 1700 – 1950.* (Barnsley: Wharncliffe Books, 2008).

Duckham, H and Duckham, B *Great pit disasters: Great Britain 1700 to the present day.* (Newton Abbot: David & Charles, 1973).

Elliott, B *South Yorkshire mining disasters, vol.1, The nineteenth century.* (Barnsley: Wharncliffe Books, 2006).

Nadin, J *Lancashire mining disasters, 1835-1910.* (Barnsley: Wharncliffe Books, 2006).

www.dmm-pitwork.org.uk/html/menu1.htm *Coal Mining Disasters*

www.senghenydd.net *Senghenydd: a history in photographs.*
Website dedicated to compiling the definitive photographic history of Senghenydd.

www.users.waitrose.com/~philipclifford/ *A brief history of mines rescue in the UK*

Year	Colliery	Deaths
1835	Wallsend, Northumberland	102
1856	Cymmer (Rhondda), Glamorgan	114
1857	Lundhill, Yorkshire	189
1860	Black Vein, Risca, Monmouthshire	142
1862	New Hartley, Northumberland	204
1866	The Oaks Colliery, nr. Bradford	361
1867	Ferndale, Pontypridd, Glamorgan	178
1875	Swaithe Main, Barnsley, Yorkshire	143
1877	Blantyre, Lanarkshire	207
1878	Wood Pit, Haydock, Lancashire	189
1878	Abercarn, Monmouthshire	268
1880	Black Vein, Risca, Monmouthshire	120
1880	Seaham, Co. Durham	164
1880	Naval steam Coal, Glamorgan	101
1885	Clifton Hall, Lancashire	178
1890	Llanerch, Monmouthshire	176
1892	North's Park Slip, Glamorgan	112
1893	Combs, Dewsbury, Yorkshire	139
1894	Albion Pit, Glamorgan	290
1905	National, Wattstown, Glamorgan	119
1909	West Stanley, Co. Durham	168
1910	Wellington, Whitehaven, Cumberland	136
1910	Pretoria Pit, Hulton, Lancashire	344
1913	Universal, Senghenydd, Glamorgan	439
1918	Minnie Pit, Podmore Hall, Staffs.	155
1934	Gresford, nr. Wrexham, Denbighshire	265
1947	William Pit, Whitehaven, Cumberland	104
		5,107
1966	Aberfan, Glamorgan	144

Figure 1 - Mining disasters with more than 100 fatalities.

Unions

As did many other trade unions, those representing miners came into being because of the terrible conditions which existed at the dawn of the Industrial Revolution, and into the first quarter of the nineteenth century[19]. At the end of the eighteenth century Scottish miners organised themselves into temporary combinations to protest against conditions, wages and other grievances. *The United Association of Colliers* was established in the North East in 1825 and published its 'Rules and Regulations'[20], together with a classic list of grievances in the same year [21]. This might be described as the first true miners' union, and at first had great success under the leadership of Tommy Hepburn.

In 1841 the first national union, *The Miners' Association of Great Britain and Ireland*, was founded in Wakefield, under the leadership of Martin Jude[22]. This was a federation of county unions which, by 1844, had a membership of over 100,000 men[23], although this too was a short-lived organisation. After giving its support to the men of the North East in their strike of 1844 it had lost most of its members by 1848, and disbanded by 1855[24]. 1858 saw the formation of the *South Yorkshire* and the *West Yorkshire Miners' Associations*, and in 1863 *The Miners' National Union* (or *Association*) was set up under Alexander Macdonald as a loose federation of district associations[25]. Macdonald and his fellow M.P. Thomas Burt saw legislation, not strike action, as the way forward, and their success was marked by, among other things, the introduction of the Coal Mines Regulation Acts of 1860 and 1872[26]. In 1869 the rival *Amalgamated Association of Miners*, a more militant body, was established. The two unions came together in 1875.

The 1870s were important in the history of miners and their unions. In 1872 the Durham Miners' Association secured the abolition of the Bond; and in the same decade they, and many other districts including Northumberland, South Wales, Lanarkshire and parts of Yorkshire, accepted the principle of the 'sliding scale' of wage calculation, which provided for automatic wage increases and decreases as selling prices rose and fell. *The Miners' Federation of Great Britain* (MFGB) was formed in 1888. It was totally opposed to the 'sliding scale' so at first those areas which supported it were not admitted to membership. It was not until 1908, after the Eight Hours Day Act was passed, that the final county association joined the MFGB. As its former General Secretary has written, 'The miners' organisation then entered upon a new phase ... It began the work of establishing national agreements with the employers as a whole[28].

In January 1919 the Federation demanded a substantial wage increase, a six-hour day, and nationalisation. To avoid a strike, the Sankey Commission was established and recommended a seven-hour day, and wage increases of 2s. per shift[29]. However in 1921, when the export market for coal collapsed, wage cuts were enforced after a three-month lockout. Market conditions improved slightly in 1923, but deteriorated again by 1925, when its membership had reached some 774,000[30]. The coalowners proposed a cut in wages and in hours and this led to the famous demand of A.J. Cook of the MFGB: 'Not a penny off the pay, not a minute on the day'. The government provided a subsidy and appointed another Royal Commission under Sir Herbert Samuel. It recommended against nationalisation, and the subsidy; and for a reduction in working hours and/or wages[31].

On 1 May 1926, the day after the subsidy ended, the General Strike began. After only nine days it was called off, leaving the miners to struggle on alone for a further seven months [32]. The strike bankrupted the MFGB and saw the beginnings of 'company' or 'non-political' unionism with the establishment by George Spencer of the *Nottinghamshire Miners Industrial Union* [33].

After several years of powerlessness, the Federation was able to exert more influence after the demand for coal increased after 1935. In 1942 'a new and strategically important Ministry of Fuel and Power' was established. National negotiations again took place, a national minimum wage for the industry was agreed and, in 1943 a new National Conciliation Board established[34, 35]. This had been a major demand of the MFGB. and its success led to the creation of the *National Union of Mineworkers* (NUM) on 1 January 1945, with a membership of 533,000. Two years later the National Coal Board (NCB) came into existence.

The next thirty years saw huge changes in markets at home and abroad, with competition from oil and gas. Economies were achieved by closing collieries, but investing in those with some future, opening up some new fields, and by holding down wages. The numbers employed, and union membership, decreased significantly. In January 1972 the first national coal strike since 1926 began.

The NCB continued to make huge losses and the industry was in a downward spiral. A Monopolies and Mergers Commission report inquiry into the Coal Board in 1982 stated plainly that it was an uneconomic industry and that pits should close [36]. This brought immediate condemnation from, and confrontation with, the NUM. A strike was called, which was to last 362 days from 5 March 1984 to 3 March 1985. It was a physical confrontation from the beginning with no quarter given on either side.

After a long and bitter dispute the decision to return to work, when it came, was not overwhelming, ninety-eight votes to ninety-one. By this time the union was split and terribly weakened, losing half its membership, with a rival *Union of Democratic Miners* enjoying fair support. In the end the government got its way and from 1984 to 1987 the number of colliery workers fell from 181,000 to 108,000[37].

 www.num.org.uk *National Union of Mineworkers*

Strikes

Strikes in coalmining have always been common, and out of all proportion to those in other industries. By the 1950s 70% of all strikes in the UK were in mining and quarrying [38], and there were national strikes in 1926, 1972 and 1984/5.

Various reasons have been put forward as to why this 'propensity to strike' exists. However, there is little doubt that the Samuel Commission of 1926 was correct when it reported that miners:

'are well aware of the history of their industry. Through reading and through family tradition, abuses which have been remedied and which may have been forgotten by others are kept alive in the memory'[39].

 Arnot, R P *The General Strike May 1926: its origin & history.* (Wakefield: EP Publishing, 1975). (originally published Labour Research Department, 1926). A compilation from contemporary sources by the respected mining historian.

 Church, R and Outram, Q *Strikes & solidarity: coalfield conflict in Britain 1889-1966.* (Cambridge: Cambridge University Press, 1998).

Banners

Sid Chaplin described the raising of the banner of his local lodge as 'the loveliest sight in all creation when, with a sharp flap, the picture billowed and sailed away'[40]; and to Norman Willis, General Secretary of the T.U.C., they were 'the most vivid and colourful part of our heritage'[41].

It is possible to trace the history of trade union banners back to the coats of arms of the ancient trade guilds; with the earliest union banners probably being produced from the beginning of the nineteenth century. The mass meetings of miners in the Durham coalfield in the 1830s were accompanied by numerous banners 'of the gayest description, nearly all being embellished with a painted design, and with a motto ...' [42]. In the North East the annual Northumberland Miners' Picnic (from 1866) and the Durham Miners' Gala (from 1871) were the chief opportunities to march behind and display banners. A handbill for the 110th Gala in 1994 carried the message 'Let us demonstrate by our presence at this great event that although they have destroyed our industry they cannot destroy our traditions' [43]. These traditions include the use of black drapes on a lodge banner when a member had lost his life in the preceding year. Sometimes banners were carried at funerals, especially when there were a large number of deaths in a disaster. May Day rallies and other Labour gatherings were other occasions when banners were unfurled and displayed with pride. At the end of the 1984/5 strike many lodge members returned to work *en masse* behind their banner, as a final demonstration of their unity.

The themes depicted on banners range from unity and fraternity to the struggle for better and safer working conditions, housing, education and healthcare. Many include portraits of leaders and Labour politicians, others feature mottoes and illustrations of a religious nature. Some of the more recent banners, painted during and after the 1984/5 strike are perhaps more overtly political.

Plate VI - Banner of the N.U.M. North East Area.

35

Union lodges all over the country used their banner as a symbol of the strength, in most coalfields, from South Wales to Scotland. Examples of those which remain can be seen at a number of locations. There are thirty nine banners in the South Wales Miners' Library. Twenty seven of them are from NUM. (South Wales Area) lodges, while the other twelve are from a number of different groups. Some can be viewed on the website of the South Wales Coalfield Collection. Many museums including the People's History Museum, the National Coal Mining Museum for England, Beamish, and Woodhorn all have collections of banners. Others may be found in miners' welfare halls, churches, chapels, community centres, council offices, libraries, in fact in any public building in a former coalfield community. The East Durham & Houghall Community College in Co. Durham has a project of preservation and restoration of banners and in 1999 NUM. (Durham Area) and TUPS Books began publication of a new journal *Bands and Banners*.

 Emery, N *Banners of the Durham coalfield*. (Stroud: Sutton Publishing, 1998).

 www.swan.ac.uk/swcc/

CHAPTER 4
How did he live?

Communities

Before the eighteenth century, in most parts of the country except the North East, coalmining was almost a branch of agriculture as it was carried out by labourers on an estate which happened to over-lie coal measures. The miners themselves were part-time, with other duties on the land, and lived with their families in the traditional tied-cottage. Later, in the Midlands, miners continued to live near workers in other industries, but in Scotland and the North East isolated mining villages began to be built[1].

The traditional nineteenth century mining community is synonymous with all that is dreary and depressing. In our minds we picture a remote village or small town, with row upon row of two-storey terraces, or lines of pit cottages, a few grimy shops, several chapels, more than several pubs, and a barrack-like school. Mining communities like this could be found in most coalfields, from South Wales to South Yorkshire and from Lancashire to Lanarkshire, and have been vividly described by many who lived in such communities. Modern historians have pointed out that, although there were many examples, these mining communities were not entirely typical. In the

Black Country, the Rhondda, and South Yorkshire for instance, many miners also lived in larger towns or cities. In Lancashire by 1911, more than a third of all miners lived in Manchester, Burnley, Oldham and the large towns of the county.

Plate VII - Colliery Dwellings.

There were even collieries in the middle of towns such as Bradford Colliery in Manchester[2] and Wearmouth Colliery in Sunderland, just over the river from the city centre and now a venue for Premier League football at the Stadium of Light.

However, it is the traditional mining community – village or small town – on which we will now focus. In fact, these words may be a misnomer. Sid Chaplin, writer and novelist, and who was born and brought up in one, described them as 'camps' which were thrown up overnight and never meant to house miners and their families for generations [3]. They were certainly built speedily, as was reported in 1842:

'Within the last ten or twelve years an entirely new population has been produced. Where formerly there was not a single hut of a shepherd, the lofty steam-engine chimneys of a colliery now send their volumes of smoke into the sky, and in the vicinity a town is called, as if by enchantment, into immediate existence'[4].

The following description of housing in Merthyr in 1848 serves to provide a general picture of most mining communities throughout the country, at this time and in some cases for many years afterwards:

'The interior of the houses is, on the whole, clean ... It is those comforts which only a governing body can bestow that are here totally absent. The footways are seldom flagged ; the streets are ill-paved, and with bad materials, and are not lighted. The drainage is very imperfect ; there are few underground sewers, no house drains, and the open gutters are not regularly cleaned out. Dust bins and similar receptacles for filth are unknown ; the refuse is thrown into the streets. Bombay itself, reputed to be the filthiest town under British sway, is scarcely worse! The houses are badly built, and planned without any regard to the comfort of the tenants, whole families being frequently lodged – sometimes sixteen in number – in one chamber, sleeping there indiscriminately. The sill of the door is often laid level with the road, subjecting the floor to the incursions of the mountain streams that scour the streets. The supply of water is deficient, and the evils of drought are occasionally felt. The colliers are much disposed to be clean, and are careful to wash themselves in the river, but there are no baths, or wash-houses, or even water pipes'[5].

In the nineteenth century many miners in the North East, and in Scotland, lived in houses provided at little or no rent by the employer. They had the virtue of being convenient for work, but had little else to commend them. Providing living accommodation also gave the employer considerable leverage in any disputes with his workforce. The threat of losing one's home was often sufficient to force a striking miner back to work, although there were many examples of evictions. The location of a mining community depended only on the whereabouts of the colliery; consideration of physical features, communication, or proximity to towns did not apply. As the Royal Commission on the Coal Industry of 1925 reported:

'... the houses must be placed where the undertakings are, and the undertakings must be placed where coal is situated'[6].

This led to the remoteness of many communities, the cause, many believe, of the miner's independence of spirit and his conservative nature. The coming of the railways provided better opportunities for travel and communication, and certainly mining families appear to have had little problem in moving house as new pits opened or conditions improved. In the 1870s fourteen new collieries were opened within a two-mile radius of Hamilton West station in Lanarkshire[7].

Another result of this remoteness was the widespread use by employers of the 'Tommy Shop' or 'truck'. This system took two forms – payment in groceries rather than cash and, more common in the coalfields, the establishment of a 'company store' and payment in tokens or vouchers, which could be exchanged for goods only at this store. South Wales, the West of Scotland, and the Black Country, where it lasted for many years, were the main areas and in the 1860s more than 40% of miners were 'paid' in this way. There is an illustration of a truck token on the website of the National Mining Memorabilia Association.

 http://www.mining-memorabilia.co.uk/Tokens.htm

After two years of pit brow work in South Wales, Janet Jones told a Tredegar court in 1864:

> 'My wages have been 5s. 6d. a week. During the whole time, I did not receive any money. I never received a farthing in money, but all the goods at the shop. The whole of my wages were swallowed up in shop bills for bread and tea, on which I lived'[8].

The iniquities of the truck system led to the establishment of the coop-erative store in many mining communities, particularly in the North East, South Yorkshire, and South Wales. Often there was opposition from the owners and agents, but the miners persisted, and the arrival of the co-op was an indication that the community was becoming more established [9]. In later years co-operative societies would provide an ever-increasing range of services, including housing for rent.

A study in the 1920s revealed that little progress had been made in improving the living conditions of miners and their families. Included in it are some shocking photographs which vividly portray the housing conditions which still prevailed, and which could very easily have been living conditions fifty to seventy five years earlier[10].

Plate VIII - Mining community, 1920s.

Model villages were built by some of the more enlightened colliery owners, such as Maltby at Bolsover, Woodlands near Doncaster[11] and the garden village built by Powell Duffryn in the Rhymney Valley[12]. Generally, however, the mining village was still an ugly and depressing place. J.B. Priestley certainly thought so in 1934:

'Imagine then a village consisting of a few shops, a public house, and a clutter of dirty little houses, all at the base of what looked at first like an active volcano. This volcano was the notorious Shotton 'tip', literally a man-made smoking hill ... [It] towered to the sky and its vast dark bulk, steaming and smoking at various levels, blotted out all the landscape at the back of the village'[13].

The beginning of the twentieth century saw the birth of the 'garden city' movement which, before the First World War, spread to mining communities. Garden Villages were developed in the South Yorkshire and South Wales colliery districts with provision for ample garden spaces, recreation grounds, tree-lined streets, and social institutes, with 'every care being taken to foster the co-operative spirit and to make the villages pleasant communities' [14]. The Dukeries coalfield of Nottinghamshire, which was not developed until the 1920s, contained many model villages such as Ollerton and Blidworth, which was built on 'garden suburb principles' [15]. The new towns movement attempted to harness the garden city concept as an aid to planning and development after the Second World War, and one of the towns – Peterlee – named after a great miners' leader – was built to solve the problems of housing and social deprivation in the mining villages of

East Durham. The descriptions of the deplorable conditions in these villages in the 1940s, as described by the planners, certainly make familiar reading and indicate that very little progress had been made in the previous one hundred years.

Regardless of the deplorable conditions outside, once the miner's front door was closed behind him it was a different world. Commentators from the earliest days of mining, such as Howitt in 1842, have described, often with some amazement, the fine furniture and contents of the miner's home:

> '... their chests of drawers, each with a japan tea-tray reared upon it against the wall; their clocks, good chairs, corner cupboards, and shelves of crockery, some of them even pieces of carpet, and all seemed to pride themselves on a good four-post bedstead with mahogany posts and chintz hangings'[16].

One cannot generalise about mining communities, they are as 'collectively varied and individually complex as human beings'; so they too 'pass through youth to maturity, on to old age and death' [17]. Unfortunately, in the first decade of the twentieth-first century, the vast majority of them no longer have at their centre their original *raison d'être* – a working colliery.

Families and family life

There have been few other industries where the work which a man did had such an impact on the way in which he and his family lived. In the early days the miner had to reside close to the mine, probably in an area remote from other settlements; so had little choice but to live in a house provided by his employer. He started work early, and worked shifts, as did his sons and his lodgers. So men were coming and going at all times of the day and night, all requiring a hot meal on the table and a hot bath in front of a roaring fire. As Sid Chaplin says 'The very nature of pit work made most women slaves, wives and daughters all' [18]. For it was the women of the family on whom the burden of cooking and cleaning, washing and baking, fell. 'The duties of a pitman's wife are very numerous', wrote John Roby Leifchild in 1853, yet he also comments favourably on the neatness and order of their homes [19]. Other writers at this period also admired the fine furniture and generally well-kept interiors of the miner's cottage, all of which were a tribute the hard work of the miner's wife.

Of course such work was not unfamiliar to her. Very often she was the daughter of a miner (as her husband was a miner's son) and she was well aware of her domestic responsibilities, from observing and helping her own mother from an early age, as reported by the *Cornhill Magazine* in 1862:

'she cherishes all the old associations of a similar home, and what constituted her mother's pride stimulates hers'[20].

Although some miners' daughters went into service, the majority stayed at home until marriage to a local lad, 'whole villages are thus often related by marriage'[21]. She then went on to have several children – large families being common in mining communities (see Chapter 2).

The education of these families was not seen as a particularly high priority and good quality teaching was in short supply until education became compulsory after 1870. Boys often started work at the age of seven or eight, and were too exhausted to attend any form of 'night school', although for a few years at least, attendance at Sunday School was quite common. It is no surprise then that the Children's Employment Commission's *Report* of 1842 described the educational state of children in the mining districts as 'disgraceful to a Christian country'[22] and described the provision of a new nursery school in a colliery village as 'certainly a novelty'[23]. Some coalowners had attempted to provide and/or support schools in mining areas, as a means of exerting influence on the pupils, and their fathers. Schools were seen by the land- and coal- owners, magistrates, and clergy as a stabilising influence on a shifting and feckless population; but in the majority of cases the miner saw them as unimportant and unnecessary, for himself or his children.

Life stories

One of the family historian's most longed-for discoveries is a biography or autobiography of a member of the family. This could be a diary (which may have been published and analysed, such as that of North East mining engineer Matthias Dunn)[24]. There could also be recorded and/or transcribed reminiscences deposited in a local archive[25], and used by others to illuminate a regional or mining history. A number of oral histories of miners have been published as *Essays in oral history* in the *Bulletin* of the Society for the Study of Labour History[26].

Other possibilities include published biographies or autobiographies. Formerly, these were more likely to be the lives of miners' leaders or of politicians or coal-owners, rather than the ordinary pitman, and mining histories seem rarely to concern themselves with the rank-and-file, just the executives[27]. John Burnett has written about the autobiographies of working people and his *Useful Toil* includes extracts from the autobiographies of three miners: B.L. Coombes (b. 1894) from South Wales, Thomas Jordan (b. 1892) from Durham and Emanuel Lovekin (1820-1905), a 'butty' from Staffordshire[28]. Coombes has actually published three volumes of autobiography. Recent years have seen the proliferation of small publishers, or self-publishing, which has opened the door to many men and women from mining communities to tell their stories in print.

You could also experience the same surprise as the present writer, when he found himself reading about the exploits of one of his own ancestors in the autobiography of a famous mining engineer, with whom he had visited South Africa in 1891 in search of coal[29].

Over eighty published autobiographies of miners or their families have been recorded, although is believed there are others still to be identified, and more to be written. They often tell of the dangers of working underground, the suffering of families after disasters, the harshness of pit life, and possibly the author's escape by means of chapel, union or politics, to a wider world. The earlier accounts will tell of childhood initiation into a man's world at a very early age[30]. Invariably the writers see themselves as spokesmen for, and their lives as representing, their colleagues as much as themselves; and they are often addressed to fellow-miners[31].

Often personal details are omitted as being of little interest, so should an autobiography have been written by a member of your family, you may not actually learn much about your forebears. As pictures of mining life over the last two hundred years, however, autobiographies are invaluable, as long as one is aware of the inherent bias and lack of objectivity. There are examples from every period, such as Bill Williams' diary of the 1870s, Edward Rymer's 'sixty year struggle for life' from the 1890s, and the life of Tommy Turnbull, who died in 1982[32].

To track down these and other life stories one can consult various guides[33], and/or visit an appropriate archive or library.

 Benson, J *British coalminers in the nineteenth century: a social history.* (Dublin: Gill and Macmillan, 1980). Chapter 4, pp. 81-111.

Many miners and ex-miners have taken advantage of the Internet and included their reminiscences of life and work in the coalmining industry on personal or mining history websites, e.g.:

 www.brocross.com/poynton/intro.htm

 www.jnadin1.50megs.com/

 www.users.waitrose.com/~philipclifford/

CHAPTER 5

Was he Church or Chapel?

John Wesley, the founder and first leader of Methodism, always had an affinity with miners and their families. In April 1739 he preached, with much success, his first open-air sermon to the colliers of Kingswood, near Bristol. Previously his colleague George Whitefield had preached there to those who 'feared not God and regarded not man'[1], and attracted great crowds, being gratified to see 'the white gutters made by the tears, which plentifully fell down their black cheeks'[2].

He also visited the Somerset coalfield on numerous occasions, often preaching in the open air, to start and support new societies. His reception in the early days was mixed. At Shepton Mallet he was forced to take refuge by mobs, but at Coleford he was received by 'honest colliers' who prayed with him in the evening and again, before he left, at 5.00am. Wesley referred to the Coleford Society as his 'second Kingswood' and visited it some twenty-two times before his death[3].

At first the miners were no more inclined than anyone else to accept Wesley's preaching, and many of them took part in the threats against him organised, in many cases, by magistrates and the Anglican clergy. One of the worst outbreaks of violence was at Wednesbury in 1743, when a mob

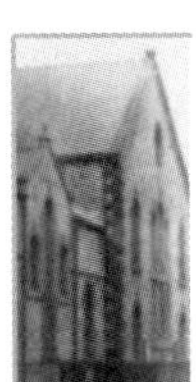

of colliers attacked the homes of known Methodists and destroyed all their furniture[4]. In other areas, Wesley met with a better reception – notably in Newcastle upon Tyne, the scene of his first preaching in the North. It was the first of many visits, and it was said that colliers would come into Newcastle and sleep in the open to hear him preach the following morning. Chowdene, a mining village near Gateshead, became Wesley's 'Kingswood of the North'[5]. On Easter Monday 1743 he preached at another pit village in Northumberland and was able to write of it in his *Journal* a few years later:

> 'The Society of colliers here may be a pattern to all societies in England. No person ever misses his class; they have no jar of any kind among them; but with one heart and one mind provoke one another to love and good works'[6].

By 1768 Wesley could write:

> 'no Indians are more savage than were the colliers of Kingswood, many of whom are now an humane, hospitable people full of love to God and man; quiet, diligent in business; in every state content; every way adoring the Gospel of God their Saviour'[7].

John Wesley died in 1791, and Methodism gradually moved from being a society of his followers within the Church of England to a 'connexion', and later a church in its own right. In fact there were soon several churches, with little difference in beliefs but often considerable variation in administration and ethos. The Wesleyan Methodists continued to be the largest grouping and expanded throughout England. However it was the Primitive Methodist connexion that appealed more to the miner. As one historian of Primitive Methodism has written:

> 'Methodism found the miners in serfdom. The breath of heaven, operating through that agency, awakened their manhood. They discovered themselves and their heritage, and altered their environment. Domestic decency and happiness, social purity and progress, industrial equity, educational expansion, political freedom, and national righteousness have been stimulated by Primitive Methodist propagandism, whether those affected by its agents have allied themselves with it or not'[8].

As the nineteenth century wore on, the influence of Methodism was to be felt throughout the country – the North East, parts of Lancashire, Yorkshire, and the East and West Midlands; and South Wales. In Staffordshire in the 1830s prayer meetings were held underground after the midday break; in Warwickshire the miners loved the 'gospel sermons of the Ranters' (Primitive Methodists); one of the earliest recorded public meetings of miners in Chesterfield began with singing and prayer[9]. Such success was partly attributed to opening of new collieries and the influx of members from elsewhere, but was also ascribed to 'the mighty outpouring of the Holy Spirit upon the unconverted and impenitent'[10]. In their classic history of *The Town Labourer*, the Hammonds suggest that it was the dangers of underground employment that prompted miners to seek 'the special and miraculous sense of protection' which Methodism provided[11].

At first, services and meetings were held in the open air, or in the homes of the miners. But very soon a chapel was needed, and this was financed, and often built, by local members. Often the land might be given by the landowner (who probably owned the colliery, the miners' homes and the rest of the village too). One of them said, when applied to for aid towards the creation of a chapel, 'O yes, I will help you, for your preachers have done so much good amongst our men that we have much less to subscribe for policemen, and for trials for misconduct'[12].

Other nonconformist denominations also prospered in the coalfields, for:

> 'As soon as new works are opened, and the cottages around them begin to be inhabited, or as soon as the population around old works increases, Dissenting chapels of the Wesleyans, Independents, Baptists and Primitive Methodists spring up, and endeavour to attach to themselves congregations'[13].

Plate IX - Chapel.

Often the chapel that the family attended depended on the work done or the seniority of the miner. Wesleyan Methodism was more middle-class, and a spiritual home for the overman or viewer rather than the hewer. Before Methodist union in 1932, and in many cases for several years afterwards, many mining communities included two or more Methodist chapels. In the Nottinghamshire mining town of Eastwood, around the turn of the century, the colliery officials and the butties (sub-contractors) attended the Congregational chapel, whereas the Baptist church was mainly supported by miners and their families [14]. If he lived locally, the mine owner, if a churchgoer at all, would almost inevitably attend the Parish Church. Support for church, or chapel was not universal, however. In the middle of the nineteenth century it was reported that in Bilston in Staffordshire four thousand out of five thousand miners in the district did not attend a place of worship of any kind[15].

In Wales in 1867 it was estimated that there were 3,107 nonconformist chapels – broadly speaking Calvinistic Methodist in the north and the Independents and Baptists in the south[16]. One reason for the popularity of nonconformity in South Wales was its opposition to the coal- and land-owners, who were often English and Anglican[17]. However, by the beginning of the twentieth century the chapels' insistence on a biblical Sabbath and teetotalism, and antagonism to trade unionism, led to a decline in membership.

In 1841 *The Scotsman* reported that 'hundreds of grown-up colliers ... never enter a place of worship, but spent the Sabbath in the vilest debauchery and rioting' [18]. Indeed the influence of church or chapel does not appear to have been particularly strong on the Scottish miner at any time in the eighteenth or nineteenth centuries. It has been argued that, in Scotland, religion was not as powerful a force in the social life of the mining community as elsewhere[19]. The miner's serfdom seemed of little interest to the Church of Scotland which, like the Church of England was seen by the miner as being allied with the coal owner - 'the kirk session care nought about us' was the judgement of one old miner[20].

There were, however, a large number of Roman Catholics in West and Central Scotland, as well as in the Lancashire coalfield. Many had arrived from Ireland, especially after the famine and epidemics of the 1820s and the 'hungry forties'. They were an 'important – though often overlooked – minority' [21] who often lived apart from their Protestant neighbours in the mining communities and were subject to much prejudice throughout the century.

The Church of England had very little support in mining areas, so much so that in many of them it was Methodism that might be considered the established church. Although Anglican clergy might deplore the behaviour of miners, and the conditions under which they were forced to exist, it took many years before they attempted to tackle the problem and build new churches to serve the mining communities. Even so, in 1882 the Archdeacon of Durham was still lamenting the miners' 'prejudice against black coats' and the clergy's 'prejudice against black faces', adding 'you need not be afraid of coal dust. It will wash off'[22]. Although writing in 1910, the Bishop of Southwell could have been describing any coalfield, at any time in the previous century:

'Year by year there is new development. Country villages change suddenly into colliery villages, where churches or mission rooms are needed. In such parishes often times the staff of the clergy is wholly inadequate, and the mass of the people, ignorant and unshepherded, are tempted to drift in to the ranks of the indifferent or hostile, save where other religious bodies assent their influence'[23].

It was to be nearly a century later before another, newly appointed, Bishop could speak of his personal memories of coalfield life:

'I can see the miners now – those were the days before pithead baths – their faces blackened, their clothes dusty, but their sparkling eyes and red lips were symbols of the vigour and vitality, the humour and the humanity that lay beneath the coal dust. I think specially of my own old Sunday school teacher in whose hands, and on whose face, the coal had left permanent marks and disfigurement. But nothing disfigured his faith and conviction'[24].

Although the influence of 'chapel' on the mining community was as extensive as it was varied, it was perhaps not as overwhelming as some partisan historians have described. It did, however, contribute to changes in society that can still be seen in the twenty-first century. Such changes were not the *raison d'être* of Methodism, but conversion inevitably brought about a change of life to one of earnestness, sobriety and industriousness. The effectiveness of Methodist missionaries was, in 1880, described thus:

'Deep emotions, loud responses, and sometimes faintings and convulsions attended the preaching and other religious services among the pitmen. Not infrequently persons were so powerfully wrought upon that they could not stand, but fell to the ground, or fainted on their seats. But the genuineness of the work was proved by its fruits. A general reformation of manners was witnessed. Sobriety, industry, and peaceable behaviour took the place of drunkenness, indolence, brawls, and contentions. Masters of collieries could not but observe the change'[25].

This led directly to improvements in community values, stability and re-assurance – and the provision of community services – in a time of insecurity and change[26]. It also provided the miner with skills and experience which he was able to use on a wider stage – public speaking, the conduct of meetings, the marshalling of arguments, committee work and so on.

Labour leaders, such as Herbert Morrison and Morgan Phillips, have suggested that trade unionism and the Labour Party itself owe more to Methodism than

Marxism[27], although this view would have surprised the movement's founder – for John Wesley was a High Tory and would have had little sympathy for trade unions. However by the end of the nineteenth century many Wesleyans had become trade union leaders, such as Ben Pickard, President of the Miners' Federation of Great Britain 1889-1904, who described his early experiences:

'As I proceeded from one mining village to another and saw the destitution and impoverished condition of the people, with the children going about barefooted and barelegged up and down the little streets, I came to the conclusion that better things should be the lot of the mining population ... I made the determination to do my level best to win for the workers better conditions and fairer wages'[28].

The Primitive Methodists were more favourably inclined towards trade unionism, in fact, with so many working-class members, it often felt the adverse effects of strikes and lock-outs. During the 1844 strike in the North East frequent union meetings were held in Primitive Methodist chapels, where 'prayers were publicly offered up for the result of the strike' and miners attended 'to get their faith strengthened' i.e. to encourage each other in the confidence that the strike would succeed'[29].

Many miners' leaders, such as Thomas Burt, Alexander Macdonald and John Wilson had a Methodist background, and the tradition continued with Frank Hodges, Jack Lawson, Nye Bevan, and George Thomas.

At the spiritual level, Methodism provided the miner with hope and security in a hazardous working environment – the very valley of the shadow of death. To mining families it provided the drama of conversion; and the imagery of the hymns brought colour to their lives and enlivened a humdrum existence. It offered new opportunities outside the home – preaching, teaching and administration for women as well as men – and an alternative to the tavern. Methodism also provided self-improvement, discipline, and education. Taken over all, its influence was considerable, and mining communities would have much the poorer without it.

Those of us with coalmining ancestors who were non-conformists, of whatever denomination, should not make any assumptions about the location of their baptism, marriage or burial records. These life events may have taken place at their particular place of worship, in which case the guides listed below will prove invaluable. Alternatively they could have been in the parish church. Indeed for baptisms they could well be found in more than one register.

Additional Sources

Benson, J, *British coalminers in the nineteenth century: a social history.* (Dublin: Gill and Macmillan, 1980). pp. 165-171.

Breed, G R *My ancestors were Baptists.* rev. ed. (London: Society of Genealogists, 2007).

Clifford, D *My ancestors were Congregationalists.* rev. ed. (London: Society of Genealogists, 1997).

Gandy, M *Basic facts about tracing your Catholic ancestry in England.* (Bury: FFHS, 2002).

Gandy, M *English nonconformity for family historians.* (Birmingham: F.F.H.S., 1998).

Leary, W *My ancestors were Methodists.* 3rd. ed. (London: Society of Genealogists, 2005).

Ratcliffe, R *Basic facts about ... Methodist records for family historians.* (Bury: F.F.H.S., 2005)

CHAPTER 6

How did he spend his leisure time?

In the late eighteenth and early nineteenth centuries, for most people, leisure was almost unknown. Coalminers, and other members of the working-class, had no genuine leisure time. If they had, it would almost certainly have been frowned on by their social superiors who feared it would lead to idleness, boredom and possible social problems[1].

The most popular sports at the beginning of the nineteenth century were those that reflected the harsh conditions experienced by the coalminer. Cock-fighting, dog-fighting and prize-fighting were popular spectacles and it has been suggested that they provided an outlet for their anxieties, aggression and frustration[2].

The working conditions underground for the coalminer were always dangerous, unpleasant and uncomfortable. When he reached the surface the miner was relieved to see the sky and know that another shift was over. Jevons suggested that, before World War I, although the miner's physical strength was exhausted, his work:

'is of such a character that it cannot absorb all the energies of his mind ... Combined with the monotonous and unexciting character of the employment,

working in dark and grimy surroundings, and living in secluded districts where all are of one occupation, has a depressing effect upon the minds of the men and tends to drive them to extremes in various directions'[3].

For the majority, the greatest need was for a long cool refreshing drink at the pub or club. The pub was the place to meet your fellows, discuss the news, denounce management, and talk about the pit – 'more coal is still filled by miners in the pubs than they ever fill while they are at the pit'[4]. The pub was comfortable and not crowded and overrun with children, like home. Wages were paid out there, at least in the early days, and pay night was all too easily celebrated on the spot. Drink was a painkiller, a morale booster, a sleeping draught and a medicine [5]. The hard-working, hard-drinking coalminer soon became legendary, although in fact the consumption of alcohol and excessive drunkenness began to decline from the beginning of Victoria's reign. However it took many years before this reputation could be overthrown, and the temperance movement was very strong in mining communities, having the support of church, chapel, Salvation Army, and mine management alike. Managers soon recognised that drunkenness made poor colliers[6].

One of the means used by management to control the intemperance of their workforce was to provide taverns 'without the drink'[7], as well as clubs and institutes where alcohol was not permitted. There was also a strong tradition for the establishment of non-licensed libraries and reading rooms, funded by the owners or solely by the miners themselves. The fortunes of such institutions waxed and waned, but those who 'take keen interest in the more serious forms of recreation' [8] were always keen to support them. In 1850 the Public Libraries Act, and in 1870 the Education Act, promoted the spread of literacy throughout the country. By the end of the nineteenth century the Miners' Institute and Library was especially popular in South Wales. In 1892 the *Glamorgan Free Press* reported that the Maerdy Library:

'… was crowded. On all sides could be seen the hard-worked colliers, with the characteristic rim of coal dust on the eyelids. A glance round the room showed me immediately the interest and earnestness with which each was devouring the contents of some book or newspaper …'[9].

The Miners' Hall or the Institute might also be the base for a colliery band or a male voice choir and, before a purpose-built cinema or theatre was available, also showed films. However, in 1915 there were still:

'many mining villages, miles away from the nearest town, with no library, no evening classes, and no institute where men can see newspapers and periodical literature, or meet for social and business purposes. Only the public-house is ubiquitous, though the cinema is becoming nearly so'[10].

The Mining Industry Act of 1920 created the Welfare Scheme and a fund:

'for such purposes connected with the social well-being, recreation and conditions of living of workers in and about coal mines'[11].

'Welfare halls', sports pitches, swimming baths and many other cultural and recreational facilities were created and funded under this scheme.

The popularity of gardening and the growing of fruit and vegetables has a long history in mining communities. In 1842 one of the Sub-Commissioners reported that:

'A miner with much pleasure showed his little garden and expatiated on the beauties of his flowers, and might have competed with a Spitalfields weaver in his choice rarities'[12].

Flower and vegetable growing was popular from the 1840s and was regularly commended by Tremenheere in his *Reports on the Mining Districts*. Some forty years later, a Chief Inspector of Mines was to describe the miner as:

'a born gardener, and besides vegetables of amazing size, specially nurtured for the local flower show, there were beautiful blooms of many kinds, both hardy and otherwise ...'[13].

He is almost certainly referring to the miner's prize leeks, which often were 'of amazing size' after careful nurture during the summer months.

Other than drink, however, there were three main recreations for the miner and his family – religion (see Chapter 5), music (Chapter 9) and sport. Music halls were gaining in popularity, often at the expense of the public house and Saturday night became the occasion when the miner and his wife and family, visited venues of this sort[14].

By the 1880s many coalminers were working only a five-day week and began to use their free time to become involved in sporting activities. The full range of outdoor sports was to be found in the colliery village – team sports such as bowls, cricket, football and rugby; and more individual pursuits like athletics, cycling and quoits. Sports involving animals such as greyhound, whippet and pigeon racing were also popular. In the early period there was cock fighting and dog fighting, too.

There were also indoor games such as cards and draughts and indoor sports such as boxing. Gambling on all of these pastimes, as well as pitch and toss, was common.

All these activities were important to the miner as much in work as on strike. They provided an escape from harsh conditions underground, and when strikes brought hardships at home then gardening (or even gambling) came into its own to help out the family budget.

Additional Sources

Benson, J *British coalminers in the nineteenth century: a social history.* (Dublin:Gill and Macmillan, 1980). Chapter 6, pp.142-171.

Howard, S 'Leisure in the pit village: meaning and change', *North East Labour History Bulletin*, no. 27, 1993. pp. 14-23.

Hyde, E D *'Doos and dugs': miners' pastimes.* (Information sheet no. 2). (Scottish Mining Museum, 1986).

www.sulc.org/cjb/pdf/coalminers.doc *The development of working-class leisure in the coalmining regions of Staffordshire, 1840-1914.*

CHAPTER 7
Was he a Bevin Boy?

Ernest Bevin was Minister of Labour in the wartime government which, by 1941, was becoming increasingly concerned about the prospects of an adequate supply of coal. In May 1941 he applied the Essential Work Order to the coal industry, which prevented any more miners from leaving. Later in the year he withdrew miners from some other industries and the Forces [1]. There was still a shortfall and, in July 1943, he told the miners' annual conference that shortage of coal was one thing that could prolong the war:

> 'At the end of this coal year there won't be enough men or boys in the industry to carry it on. It is the one great difficulty in this war effort ... At the same time we are carrying out this invasion and every bit of territory we take from the enemy we have got to find coal for ... It is quite obvious that I will have to resort to some desperate remedies during the coming year. I shall have to direct young men to you'[2].

The scheme which took his name, and by which young men were 'directed' into the industry, began in December 1943, and its introduction received much attention from press and public alike. Most young men would have preferred to serve in the Forces, and a ballot was used 'to take some of the

sting out of the ill-luck of those who found themselves turned into coalminers'[3]. In December 1943, Bevin told the House of Commons:

> 'I therefore propose to resort to the most impartial method of all, that of the ballot. A draw will be made from time to time of the figures from 0 to 9 and those men whose National Service Registration Certificates happen to end with the figure or figures thus drawn by ballot will be transferred to coalmining'[4].

In all, 21,800 boys were recruited before the scheme was abolished in May 1945 when the war in Europe came to an end [5]. They received four weeks of physical training and were introduced to coalmining life. This was followed by two weeks training in the pit, for the kind of job for which the trainee was being taken on. The amount of coal got by the boys directly was small as only about 6 – 7,000 were employed at the coal face. Their importance was in taking over a number of less-important jobs, which freed-up some 11,000 miners for more vital work. The plan did not stop strikes nor did it revive the more worrying flagging levels of coal production, but it was an example of the imaginative and charismatic part played by one leader of labour, Bevin[6].

Plate X - Bevin boys at Cramlington Lamb Colliery, Northumberland, October 1944.

A number of 'Bevin Boys' have recorded in print their memories of what to most of them was a totally new experience. One of them, Ted Holloway, has left a pictorial record too, of his experiences from September 1944 to December 1947.

Holloway, G ed *A Bevin Boy remembers*. (Condicote: Henge Publications, 1993).

Hickman, T *Called up, sent down: the Bevin Boys' War*. (Stroud: The History Press, 2008). Based on the memories of several Bevin Boys and includes, as an appendix, potted biographies of a further 68 Bevin Boys.

Some records of the scheme survive. Wartime files on labour issues and Government policies in The National Archives refer to Bevin Boys, although according to Hickman, they consist only of records relating to the Midlands. There is an extensive collection of material in the Imperial War Museum, Department of Documents (Miscellaneous 2834). This includes a complete list of ballotees, optants and volunteers for the Midlands region for the period December 1943 – July 1947, as well as original and copied documents, photographs, etc. At least 24 hours' notice is required before any intended visit to the Department.

http://yourarchives.nationalarchives.gov.uk/index.php?title=Bevin_Boys

In 1995 there was somewhat belated recognition of the contribution made by Bevin Boys to the war effort. As part of the fiftieth anniversary celebrations of V.E. and V.J. Days, H.M. The Queen, the Speaker of the House of Commons, and the Prime Minister, all mentioned them in their speeches. In April 2003 three trees were planted at the National Memorial Arboretum in Staffordshire in tribute to the Bevin Boys of England, Scotland and Wales. Finally, it was announced by the Government in 2007 that a badge would be issued to all Bevin Boys, with the first badges being presented by Prime Minister Gordon Brown in March 2008.

http://news.bbc.co.uk/player/nol/newsid_7310000/newsid_7312800/73128 98.stm?bw=bb&mp=wm&asb=1&news=1&bbcws=1

The Bevin Boys Association was established in 1989, with a handful of members growing to over 1,800. It is an active association which holds an Annual National Reunion, with additional Regional Reunions held in various parts of the United Kingdom, usually at locations associated with the coalmining industry. Contact: The Bevin Boys Association, c/o Mr Warwick H Taylor, M.B.E., 'Little Barn', 1 Rundlestone Court, Poundbury, Dorchester, DT1 3TN, Dorset.

Additional Sources

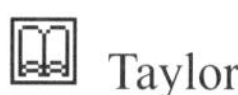 Taylor, W *The forgotten conscript: a history of the Bevin Boys*. 2nd ed. (Gainsborough: Babash Ryan, 2003). By the Archivist of the Bevin Boys Association.

 www.24hourmuseum.org.uk/museum_gfx_en/AM28467.html *The Association*

 http://seniorsnetwork.co.uk/bevinboys/bevinboysbooklet.pdf

CHAPTER 8

But my mining ancestor was a woman!

Certainly from the eighteenth century, and possibly from a much earlier date, women had worked in coalmines – mothers and daughters helping husbands and sons as part of the regular domestic economy of the day. Although by the end of the eighteenth century women were no longer employed in the coalfields of Northumberland and Durham, Gloucestershire, North Somerset, Warwickshire, Leicestershire or Staffordshire, many mines in Lancashire, North Wales and in Scotland employed a total of many hundreds of women, often underground. They were traditionally employed in haulage, although there was, in the eighteenth century, a Lancashire pit where one seam, called 'the woman's coal', was entirely worked by females[1].

The worst area for underground female labour in any terms – the proportion employed, the severity of the work, and the persistence of the custom – was the east of Scotland. Here, the haulage of coal between getter and shaft, and sometimes between getter and the surface, was almost exclusively a female occupation. As boys became getters at a very early age the proportion of girls and women in most pits was high. 'Putting' was carried out by means of the 'slype', an iron-keeled wooden box holding between two and five hundredweight, drawn by a harness which went over

the shoulders and round the waist, with the drawing thong fastened to the back of the belt and passing over the back of the legs. 'It is almost incredible', reported Sub-Commissioner Robert Franks in 1842, after seeing the slype in use, 'that human beings can submit to such employment, crawling on hands and knees, harnessed like horses, over soft, slushy floors more difficult than dragging the same weights through our lowest common sewers, and more difficult in consequence of the inclination, which is frequently one in three to one in six'.

Franks also reported on the work of an eleven year-old girl, Alison Jack, who worked at the Loanhead Colliery near Edinburgh. In 1841 she had been bearing for her father for three years. She went down with him at two in the morning and worked through until one or two in the afternoon. Her task was to carry about a ton of coal per shift.

> 'She takes her creel (a basket formed to the back, not unlike a cockleshell flattened towards the neck, so as to allow lumps of coal to rest on the back of the neck and shoulders), and pursues her journey to the wall face, or as it is called here, the room of work. She then lays down her basket, into which the coal is rolled, and it is frequently more than one man can do to lift the burden on her back. The tugs or straps are placed over the forehead, and the body bent in a semicircular form, in order to stiffen the arch. Large lumps of coal are then placed on the neck, and she then commences her journey with her burden, first hanging her lamp to the cloth crossing her head'[2].

Alison staggered about thirty yards from the face to an eighteen-foot ladder, climbed this, went along a passage, up another eighteen-foot ladder, along another passage, up more ladders, and so on until she had carried her creel to the point reached by the shaft, where she unloaded it into a skip. She reckoned to do her stint in twenty journeys; if she flagged, her father was handy with the strap.

Another woman, Margaret Hipps, reported:

> 'The pavement I drag over is wet, and I am obliged at all times to crawl on hands and feet with my bagie hung to the chain and ropes. It is sad sweating and sore fatiguing work and frequently maims the women'.

Plate XI - Margaret Hipps.

The Scottish Mining Museum has compiled two Fact Files on *Women in mining communities* and *Women and children in the mines before 1842*. These can be downloaded from:

 http://www.scottishminingmuseum.com/default.asp?tid=11

It has been estimated that, at the time of the 1841 Census, some 5,000 females were employed in the coal-mining industry – about two-thirds in Scotland, one-fifth in Lancashire and one-fifth in Yorkshire[3]. These figures were quite a shock to respectable Victorian society, and the opportunity for further investigation was presented by the enquiries of the Children's Employment Commission. This was initially directed at children and young people, adult females not being part of the terms of reference. However the practice of employing women in coalmines was condemned by the visiting Sub-Commissioners who covered the entire country collecting evidence.

They were greatly concerned about the moral risks underground to girls and young women. Whilst visiting a Yorkshire pit, Jelinger C. Symons:

'found assembled round the fire a group of men, boys and girls, some of whom were of the age of puberty, the girls as well as the boys stark naked down to the waist, their hair bound up with a tight cap, and trousers supported by their hips'.

and his difficulty in distinguishing male from female caused considerable amusement. He went on:

'One of the most disgusting sights I have ever seen, was that of young females, dressed like boys in trousers, crawling on all fours, with belts round their waists, and chains passing between their legs....' [4].

The report of the Commission was published on 7 May 1842. For the first time, a government report included engravings, which drew attention to the conditions and work which women were forced to endure underground. These engravings of women at work underground, have been reproduced in many books and periodical articles since 1842. The report includes interviews with female miners in many of the coalfields. The popular press, of course, was very keen to reproduce, and embellish, the illustrations.

The Mines and Collieries Act received the Royal Assent on 10 August 1842, but brought with it the problem of employment for the women now excluded from their former work. For many women it was the only work available and the only work they knew – so many just carried on working, and with only one Inspector for the whole country they, and their employers, were unlikely to be found out.

Neither the Census nor the Commission Reports had made a clear distinction between women working underground and those at the surface – the 'Pit Brow Lasses'. As the century progressed, the number of women working on the surface slowly declined – from 6.25% of all surface workers in 1874, to only 3.13% by 1900[5].

By 1909 the number of pit brow lasses had actually increased, and World War I saw this figure almost double to 11,300 by November 1918 [6]. Numbers then fell steadily through the twentieth century, although it was mechanisation, and then pit closures – not legislation – which brought about the end of working women in and around coal mines. Incredibly it was not until 1 July 1972 that the final two women surface workers were made redundant at Haig Colliery, Whitehaven, 130 years after legislation against women working underground.

Surprisingly, photographs of some Victorian women mineworkers still exist, due to the interests in working women of Arthur Munby – barrister and man-about-town [7]. He made a collection of photographs of pit brow girls from Wigan and South Wales and these have been described, reproduced and sometimes the subjects even identified by Michael Hiley [8].

Additional Sources

 John, A *By the sweat of their brow: women workers at Victorian coal mines.* (London: Croom Helm, 1980).
Concentrates on the pit brow lasses of Wigan in Lancashire. For a limited preview see:

http://books.google.co.uk/books?id=jyUdB9cy6QC&dq=angela+
john&source=gbs_summary_s&cad=0

John, A *Coalmining women: Victorian lives and campaigns*.
(Cambridge: Cambridge University Press, 1984). A more accessible and fully
illustrated account.

Lake, F and Preece, R *Voices from the dark*. (Yorkshire Mining Museum
Trust, 1992). Covers coalmining women in Yorkshire.

Pollard, M *The hardest work under heaven: the life and death of the
British coal miner*. (London: Hutchinson, 1984). Detailed coverage of the work
of women and children underground, based on the *Report,* and includes copies of
engravings, and later photographs of women mineworkers.

http://www.balmaiden.co.uk/index.htm
Includes the history of women in mines in Great Britain and images of Pit Brow
Lasses.

http://freepages.genealogy.rootsweb.ancestry.com/~stenhouse/coal/
pbl/coalmain.htm *Women and the pits*

www.daveweb.co.uk/pitbrowlasses.htm
Download, or order a printed copy of Dave Lane's book *Pit brow lasses*, which
brings to life those women who toiled at the pit brow of the collieries of Lancashire,
with particular emphasis on those around the Wigan area.

CHAPTER 9
The arts

In his introduction to the new edition of Galloway's *Annals*, Baron Duckham writes:

> 'Quite apart from the industry's understandable attraction for scholars, economists and social or political reformers, it has provided by turns a dramatic milieu, a subject of awe, a background of conflict or a symbol of man's elemental struggle with nature's more forbidding side, for creative writers of every kind. No other way of human breadwinning, except the call of the sea and perhaps agriculture, has exerted a similar hold on the artistic imagination,'

and he believes that this is one explanation for the fact that 'the arduous and dangerous business of getting coal has found for itself a place in the national consciousness'[1]. The purpose of this chapter is to highlight some of the many examples of that 'incongruous and unsympathetic marriage'[2] between coal and the arts.

Art

The earliest known British image with a mining association may well be that of the Ancient Free Miner as portrayed on a brass in Newland Church in the Forest of Dean (see page 2). However, it was not until the eighteenth

century that 'industrial art' began to appear, the first examples being paintings of the new landscape which was being created by the Industrial Revolution at Coalbrookdale [3]. What might be termed the 'first true mining picture' was *Pitmen at play near Newcastle upon Tyne: painted from nature*, by Henry Perlee Parker (1795-1873) and first exhibited in 1836[4].

The illustrations – which were more shocking to Victorian sensibilities than the text – in the *Report* of the Children's Employment Commission, 1842 brought a new sense of artistic realism. The same year saw the publication of the first issue of the *Illustrated London News*, which would go on to publicise many of the mining disasters throughout the coming years, and in 1844 Thomas Hair published his *Views of the Collieries of Northumberland and Durham*. All were black and white, in a style which was to find favour with Van Gogh:

> For me the English black and white artists are to art what Dickens is to literature. They have exactly the same sentiment, noble and healthy, and one always returns to them[5].

The use of photography in mines during the latter part of the nineteenth century led to the production of even more realistic images, and photographs ranged from *cartes* of Wigan pit brow lasses to postcards of mining disasters. Later Bill Brandt was able to photograph several symbolic mining scenes for his pictorial survey *The English At Home* published in 1936, such as one of 'seven Welsh miners, as black as Al Jolsons, eyes and teeth glinting, caught in a splash of sunshine as they reach the surface and wait to be released from the cage' [6].

Henry Moore was an official war artist and in 1942 spent two weeks at Wheldale Colliery, Castleford, where his father had once worked, commenting 'If one was asked to describe what Hell might be like, this would do' [7].

At the same period Graham Sutherland recorded the work of opencasting near Abergavenny, both artists 'making drawings that were within the brief of propaganda but never made subservient to it'[8]. Amateur artists also had an important role in preserving and disseminating the images of coalmining. The period before the Second World War saw the beginnings of the *Ashington Group* in Northumberland, where a group of working miners established a group which was to endure for fifty years. Their importance of artist/miners was soon recognised after nationalisation, for the NCB. first put on an 'Art by the Miner' exhibition in London in October 1947 which then, from April 1948 to March 1949, toured the country, and in the 1970s the NCB. took part in an 'Artist-in-Industry' scheme. More recently an exhibition of mining art, *When Coal was King*, was held in Kirkcaldy, Fife during 2000.

Literature

In 1985 Klaus published what he believed to be the first study of the literature of any occupational group [9]. His survey begins with Edward Chicken's *The Collier's Wedding* (1720), the earliest known work in which coalminers figure prominently. He then considers examples of miners' poetry, speeches, tracts, pamphlets, newspaper articles and letters which have survived from the nineteenth century, including Thomas Wilson's *The Pitman's Pay* (1826).

Charles Dickens, in volume 2 of his magazine *Household Words*, published 'A coal miners' evidence' concerning colliery explosions in the Black Country and the North East. This was followed-up by 'A remedy for colliery explosions' and then 'A true remedy for colliery explosions' by S.R. These articles can be viewed on the Internet Archive website:

 www.archive.org/details/householdwordswe02bradrich

There were 'pitmen novelists' too, such as Sid Chaplin and Harold Heslop, among others. One writer suggests that Heslop 'created the mining novel in Britain' then took it 'underground and challenged the reading public to follow him' [10]. George Orwell and D H Lawrence, both giants of twentieth-century literature, have described their personal experiences of coalmining life in *The Road to Wigan Pier* (1937) and *Nottingham and the Mining Countryside* (1929) [11] respectively. Other novelists have portrayed an often stereotypical view of coalmining in a variety of works over the last 150 years [12]. Two of the best known are A J Cronin's *The Stars Look Down* (1935) and Richard Llewellyn's *How Green Was My Valley* (1939). Many of Catherine Cookson's novels have a mining background, and some of these have also been filmed.

Early in his writing career, D.H. Lawrence wrote a number of plays with a coalmining background, including *A collier's Friday night* (1919). David Storey's play *In celebration*, was also filmed; but perhaps the best known play with a mining theme is *Close the Coalhouse Door* by Alan Plater which was based on some of the stories of Sid Chaplin and includes songs by Alex Glasgow. Chaplin himself described it as:

'a cunningly well-wrought play. It has great gusto and gaiety and songs that go right home. But over and above all this, it conveys the essential tragedy of his life. It is about people before miners' [13].

Music and Song

Coalminers were often members of male voice choirs and choral societies. Another popular musical activity was playing in and supporting the colliery band which, of course, played an important part in Galas and other celebrations. As early as 1827 every large colliery in the North East was said to have its own band, and many musicians went on to achieve national, and even international recognition[14].

The coalfields also gave rise to a tradition of songs, ballads and poetry and 'the songs of the pit community can be seen ... as much a part of the pitman's daily experience as the wage he hewed, the ale he drank'[15]. However, it has been argued that many of the miners' popular songs of the nineteenth century were not written by miners themselves, but for them by outsiders[16]. Of course there were 'pitmen poets', with both Joseph Skipsey and Tommy Armstrong aspiring to that title.

 Raven, J *The folklore and songs of the Black country colliers.* (Wolverhampton: Broadside, 1990).

From the 1950s onwards a new kind of poem or song began to appear – the farewell tribute to collieries which have closed, and a way-of-life which has disappeared. Many traditional, and more modern, folksongs have been recorded, perhaps the most famous by the Elliott family of Birtley.

Film

Some of the books and plays mentioned above have been filmed, and others can be located in the various film guides[17]. *Billy Elliot* has delighted cinema, and theatre, audiences, with its tale of growing-up during the Miners' Strike of 1984/5.

CHAPTER 10

The Elemore Colliery Disaster, 2 December 1886

The family historian must be aware of the importance of contemporary documents in helping to portray as full a picture as possible of the events in the life of his family. Sadly, the major source of documentary evidence about a coalmining ancestor is if he was killed or injured in an accident or explosion. Such incidents gave rise to:

- An inquest, the records of which may survive, or be incorporated in the official, or newspaper reports.

- An official report of the disaster – which will include a list of those killed, their location at the time of the disaster, their job, their age, underground plans, etc.

- The annual report of H.M. Inspector of Mines for the coalfield in question – which will include similar information to the above, as well as details of other accidents in the region. As well as reports of disasters, they are useful in that they may also include lists of all the collieries in the area; and the names (and sometimes addresses), of those involved in supervising, and taking, examinations.

- Local newspaper reports of the disaster, with illustrations, names and addresses of the dead, their dependants and so on. Later there will be a report of the funerals, and of the inquest into the deaths.

- There may be some records of the local Relief Funds set up to support the dependants.

- Some form of memorial in the parish church, churchyard, cemetery, or miners' hall is likely to have been erected. This will probably give a list of those killed.

- It is very likely that at least one description of the disaster, based on the above records and/or from the personal knowledge of the writer, has been compiled. This may well include copies of some or all of the above, and may be in the form of a book, locally published pamphlet, or newspaper or periodical article.

- Similarly, there are oral history records, on tape and in printed form, relating to such disasters.

Some of these documents, in connection with an explosion at Elemore Colliery, Easington Lane, Co. Durham, are illustrated below.

The explosion occurred at 2.55 a.m. on Thursday 2 December 1886. Forty one men were at work in the mine. The seat of the explosion was in the George Low Main area, where a roadway was being widened prior to it being lined with a brick arch. The blasting of stone at this period was very dangerous, as the only explosive available was 'black powder', which had to be ignited with a naked light. Before the explosion a hole had been drilled in the side of the roadway and charged with powder. Thomas Johnson fired the shot by opening his Davy lamp and lighting the fuse. An eyewitness described a flash issuing out of the shaft like lightning, with a cloud of dust that discoloured the covering of snow that lay around the shaft.

Twenty five men died, from the blast, or the suffocating effects of after damp (carbon dioxide). Three died later of their injuries. Ironically, Johnson, although badly injured, was able to give evidence at the inquest, evidence which contradicted the statement of John Luke, who survived for three days. Although the enquiry at the time was unable to determine the cause of the explosion, it seems likely that the disaster was a classic example of a coal-dust explosion, which occurred when the shot was fired.

This page of the 1881 Census includes three of the men who were killed – George Paterson (or Patterson/Pattison), and George Thompson and his son Robert. They lived near each other in the small coalmining community of Brickgarth, between Elemore Colliery itself and the village of Easington Lane. The Patersons are a good example of mining family which moved around the country. George was born in Edinburgh and his wife Isabella at Winlaton on Tyneside. Their children were all born in different mining communities in the North East, presumably as George, like many others, made the regular move to a new pit for better pay, or working

The undermentioned Houses are situate within the Boundaries of the

Civil Parish [or Township] of	City or Municipal Borough of	Municipal Ward of	Parliamentary Borough of	Town or Village or Hamlet of	Urban Sanitary District of	Rural Sanitary District of	Ecclesiastical Parish or District of
Hetton le Hole				Easington Lane		Hetton le Hole Houghton le Spring	Lyons & Hetton And all Angles

No. of Schedule	ROAD, STREET, &c., and No. or NAME of HOUSE	Inhabited	Uninhabited (U.), or Building (B.)	NAME and Surname of each Person	RELATION to Head of Family	CONDITION as to Marriage	Males	Females	Rank, Profession, or OCCUPATION	WHERE BORN	If (1) Deaf-and-Dumb (2) Blind (3) Imbecile or Idiot (4) Lunatic
128	Brick Garth	1		George Paterson	Head	Mar	48		Coal Miner	Edinburgh	
				Isabella do	Wife	Mar		50		Durham Winlayton	
				Elizabeth White	Daur	Widr		24		Northumberland West Moor	
				Andrew Paterson	Son	Unm	15		Coal Miner	Durham Sherriff Hill	
				Isabella do	Daur			13	Scholar	do Lumley	
				Mary Ann do	Daur			10	do	do Hetton le Hole	
179		1		Ann Collage	Head	Unm		19	Chair Woman	do Lumley	
180		1		Stephen Corner	Head	Mar	24		Colliery Labourer	do South Hetton	
				Elizabeth do	Wife	Mar		22		do Easington Lane	
				Thomas Crosby do	Son		4		Scholar	do do	
				Jane do	Daur			3		do do	
181		1		Joseph Harland	Head	Mar	28		Coal Miner	do Easington Lane	
				Elizabeth do	Wife	Mar		27		do Broomside	
				Isabella do	Daur			7	Scholar	do Easington Lane	
				Hannah do	Daur			6	do	do do	
				E. Elizabeth do	Daur			5	do	do do	
				John Wm do	Son		2			do do	
				Robert do	Son		9 Months			do do	
182		1		George Thompson	Head	Mar	44		Coal Miner	do South Hetton	
				Jane do	Wife	Mar		39		do Haswell	
				John do	Son	Unm	15		Coal Putter miner	do Easington Lane	
				Robert do	Son		12		Scholar	do do	
				Jane do	Daur			7	do	do do	
				Ann do	Daur			4	do	do do	
				Elizabeth do	Daur			3 Months		do do	
5	Total of Houses... 5					Total of Males and Females..	10	15			

NOTE.—Draw the pen through such of the words of the headings as are inappropriate.

Eng– Sheet G.

73

Plate XII - Census Enumerator's Book Brickgarth 1881. (PRO: RG 11/4976 f. 111) (Copyright waived by the Controller, HMSO).

conditions, or a bigger house. His daughter Elizabeth, aged 24, is already widowed, and his son Andrew, aged 15, already a coalminer.

From the evidence of the Census at least, George Thompson appears not to have moved around the area so much as his neighbour. He was born at South Hetton, the next village to the south of Easington Lane, and all his children were born at 'the Lane'. He is described as a coalminer, although the official report of the tragedy lists him as a hewer. His son John was a putter, and aged 19 when he died.

Brickgarth was a close-knit community of, in 1881, 213 houses in which lived 1,078 people, mostly miners and their families. The houses were built, close to the colliery, in the 1820s, at the same time as it was opened.

Although the parish church of St Michael, Lyons opened in 1869, the only burials there before the disaster were those of two of the Rector's children. There were funerals for the men killed in the explosion over three days, with various local clergy officiating. George Paterson was buried on 6 December and George and Robert Thompson on 7 December. The burial register gives little further information, and makes no mention of the disaster. Readers of later generations, not knowing the circumstances, being left to wonder why there were so many burials so close together.

All the local newspapers covered the events – from first reports, to the funerals and, later, the inquest and the publication of the official report. There was national interest too, with *The Times* giving good coverage to the tragedy and its aftermath.

BURIALS in the Parish of _Lyons_ .... in the
County of _Durham_ .... in the year One thousand
eight hundred and _86_

Name.	Abode.	When Buried.	Age.	By whom the Ceremony was performed.
1886 Robert Hills No. 17	Elemore	Dec. 6th	64 years	David Bryson, Curate
1886 Stephen Parkinson No. 18	Easington Lane	Dec 6th	27 years	David Bryson, Curate.
1886 Joseph Williams No. 19	Easington Lane	Dec 6th	38 years	David Bryson, Curate
1886 John Johnson No. 20	Easington Lane	Dec 6th	58 years	David Bryson, Curate
1886 George Askew Pattison No. 21	Easington Lane	Dec 6th	31 years	David Bryson, Curate
1886 John George Laverick No. 22	Elemore	Dec 6th	22 years	R. G. Hutt Vicar
1886 George Thompson No. 23	Easington Lane	Dec 7th	43 years	R. G. Hutt Vicar
1886 Robert Thompson No. 24	Easington Lane	Dec 7th	19 years	R. G. Hutt

Plate XIII - Burial Register, St Michael Lyons, December 1886.
(Reproduced by permission of Rev M. Beck).

AN EXPLORER'S NARRATIVE.

One of the explorers who helped in the work of recovering the bodies entombed in the Elemore Pit has given a correspondent the following narrative:—You're one of the explorers? said one of our head officials to me on Saturday afternoon, as I stood in the lamp-room at bank. I did not know that I was an explorer at that moment. I had not been down into the workings yet, but I had come in my pit clothes, ready to give assistance if I could; and when I was thus addressed, I thought I would go down. There were fully fifty of us in the "shift," and once down I was attached to the party going into the George Low Main seam. The first place we examined closely was the spot where the three stonemen, John Luke, Robert Appleby, and Henry Johnson, were engaged in enlarging the intake air-way. After a brief search we found that a cross-cut hole had been made and the shot fired. From that point we proceeded into the workings until we came to the bottom of the drift, where we examined the stoppings and found them all in. Everywhere we found indications that showed us we were travelling the way the blast of the explosion had taken, as we were also going in with the current of fresh air. But the force of the explosion apparently diminished rapidly, and ceased altogether at the foot of the drift, so that what mischief was done beyond that, must have resulted from the after-damp. On reaching the top of the landing we found the bodies of eight men. These men had apparently been trying to find their out way when they succumbed. They had all their clothes on, and some of them had their neck-wrappings over their mouths, as if for protection against the foul air. After that we came to the bank, and I joined the explorers who were going into the Low Hutton seam.

Plate XIV - Durham Chronicle. 10 December 1886. p. 7.

Rescuers were known as 'Explorers' and the newspapers carried a number of these eyewitness accounts, for instance the *Sunderland Daily Echo* of 3 December included 'A Horse-keeper's Experience' by Henry Moss, 'A Driver's Story' by Wm. Johnson, aged 14, and 'A Very Narrow Escape' by Thomas Johnson. It was many years later that the author discovered that his own great-grandfather, George Archer, had been one of these brave men.

Funeral sermons in connection with the melancholy event were preached on Sunday in nearly all the churches in Hetton, Easington Lane, Eppleton, and the neighbourhood; and special prayers were offered for those bereaved. Reference was also made to the accident in the morning in Durham Cathedral by Professor Farrar.

Six of the miners were buried on Sunday afternoon, at Easington Lane. These were—William Seeds, Robert Wilson Appleby, Ralph Lawson, Thomas Spence, Ralph Fishburn, and Thomas Robins. The interments took place in the small plot of ground before the parish church at Easington Lane. Although the ground has been consecrated, there had been no previous burial in it, and the first grave that was dug was for the man Appleby, who was the first that was removed from the mine. Nearly the whole of the inhabitants in the neighbourhood of the Hetton Collieries went in mourning, the window blinds in many of the houses were drawn, and an air of sadness and melancholy hung over the whole neighbourhood. The interest that was shown in the funerals, which took place in the afternoon, was extraordinary. Almost every house in the vicinity was emptied of its occupants, and people streamed in from Seaham, Seaton, Ryhope, Murton, Sunderland, and many other places. Before three o'clock the one long street of houses which forms Easington Lane was filled by a dense mass of people. There could not have been fewer than ten thousand people present, and these masses of people ranged themselves on both sides of the street, and crowded densely in front of the churchyard gates. A few minutes before three o'clock, the coffin containing the remains of Robert Appleby was borne on the shoulders of four men, and was met at the gates by the vicar (the Rev. R. G. Hutt) and the curate (the Rev. David Bryson), who, speaking the opening lines of the burial service, preceded the pall-bearers into the church. After a brief service, the coffin was brought outside again, and the body of the dead miner was placed in its last resting-place. The coffin, which was a plain black one, was covered with wreaths, and bore the simple inscription: "Robert Wilson Appleby, died December 2nd, 1886, aged 53 years." The bodies of the other four men and the boy Robins were interred in similar fashion. Each body was enclosed in a plain black coffin, and bore an inscription setting forth the age of the deceased and date of his death. The six graves were distinct one from the other, and were placed side by side in a row extending from the door of the church to the iron gates giving admission to the burial ground. During the burials the cemetery was packed with people, and many affecting scenes were witnessed, the mourners grieving sorely over the remains of their relatives.

Plate XV - Durham Chronicle. 10 December 1886. p. 7.

The streets of Easington Lane were thronged with thousands of people who came from surrounding villages for miles around to pay their respects. Most of the men were buried at the parish church, although two were buried at Hetton. A service for Samuel Grice was held in the Wesleyan Methodist Church before his burial with his comrades. The *Durham Chronicle* reports that he had changed shifts with a friend so that he could preach his first sermon as a local preacher. If it had not been for this, then Grice would not have been underground at the time of the explosion.

The newspaper reports also mention the setting up of a Relief Fund. Eventually this totalled £1,888. The account book of this fund has survived and is now in Durham County Record Office. It reveals that the 'Hetton, Eppleton and Elemore Collieries Accident Fund' paid out £27 9s. 10d. to the dependants of each man.

—— Richard Westman, Run Over by some laden
Coal Waggons, near Seaham Harbour, 31 *d* 5 *f*
—— Lady Williams and Miss Chichester, from
the Carriage being Overturned, both Seriously
Injured, near Barnstaple, 6 *a* 7 *a*
ACCIDENTS—MINING:—
—— at the Albion Colliery, near Pontypridd,
Four Men Killed, 5 *n* 9 *f*—6 *n* 8 *b*
—— at the Aldwarke Main Colliery, Two Persons Killed by Falling from the Cage 23 *n* 6 *b*
—— at Cwmpennar, near Mountain Ash, by a
Collision of Trucks, 11 *d* 9 *e*—14 *d* 11 *c*
—— at Elemore Colliery, Hetton, near Sunderland, Great Loss of Life, 3 *d* 10 *a*—4 *d* 8 *b*—6
d 10 *f*—7 *d* 11 *e*—8 *d* 10 *f*—9 *d* 11 *f*—13 *d* 6 *f*
—— Explosion in the Silkstone Pit, Altofts, near
Wakefield, 18 Men Killed, 4 *a* 7 *d*—5 *a* 12 *c*—
6 *o* 10 *f*—9 *a* 10 *d*—12 *a* 6 *f*—26 *o* 7 *a*
—— Relief Fund, 9 *o* 10 *d*—12 *o* 6 *e*
—— of Gas in the Conyngham Colliery, twenty
five Miners Injured, 27 *n* 5 *c*
—— at Fishby Colliery, near Walsall, 14 *d* 6 *f*
—— at Moira, Leicestershire, 9 *d* 7 *a*—18 *d* 7 *f*
—— at Snydall Hall Pit, West Houghton, Two
Men Killed, 23 *d* 7 *f*

Plate XVI - Palmer's Index to 'The Times' Newspaper, Oct 1 to Dec 31, 1886.

The index to *The Times* can be used to trace local, as well as national events. Thirteen Mining Accidents are listed in the last three months of 1886 alone. *The Times Archive* is a valuable online resource which can also be used to retrieve such information. Access is by subscription on a daily, monthly or annual basis.

 http://archive.timesonline.co.uk/tol/archive/

FATAL COLLIERY EXPLOSION.

At five minutes to 3 o'clock yesterday morning a serious explosion took place at Elemore Colliery, situate about eight miles from Durham, and one of a group owned by the Hetton Coal Company. The colliery has been worked for a period of 45 years, and, with the exception of one or two minor accidents, has enjoyed an immunity from disasters of a more serious character. Twenty-six years ago an explosion occurred at Hetton pit, about a mile distant and connected with Elemore, at which 23 lives were lost. The mine is 257 feet deep, is worked by the ordinary upcast and downcast shaft, and is lighted by the electric light. The seams of coal worked are the Lady Main, partly, the Low Main, extensively, and the Hutton seam, partly. The latter seam of coal is nearly worked out.

At the time of the disaster 38 men were in the mine. Had the explosion occurred a few minutes later the loss of life would have been greater, as a number of deputies were waiting at the bank ready to descend to inspect the mine before the day-shift men went down to their work.

The first indication of the disaster was a loud rumbling sound, which was heard at a considerable distance from the pit. On reaching the pit mouth it was at once discovered not only that an explosion had taken place, but also that the pit had fired. Information was at once sent to Mr. T. Lishman, the certificated manager, and to the other officials. An exploring party, guided by Mr. Lishman, Mr. Johnson, and Mr. Todd was at once organized. The cage being useless, they were compelled to descend in the hoop and kibble. The first man they found on reaching the Low Main seam was Thomas Charlton, whose statement will be found below. They made their way to where the shot would be fired and travelled down the return airway to the Lady Low seam. The dead body of Thos. Spence, furnaceman, was lying near the ventilator furnace, also that of Ralph Lawson, whose occupation was to watch the electric light. At the end of the shaft Appleby, Luke, and Johnson, who had been enlarging the in-take air-way, were found lying, each of them being very badly burnt. They were at once sent to bank, but Appleby died almost immediately and Luke was dreadfully burnt. His statement will be found below. It was not advisable to venture further into the workings, and the exploring party returned to the bottom of the shaft.

Plate XVII. The Times, 3 December 1886. p. 10.

Several lengthy reports of the explosion appeared in *The Times* during December 1886.

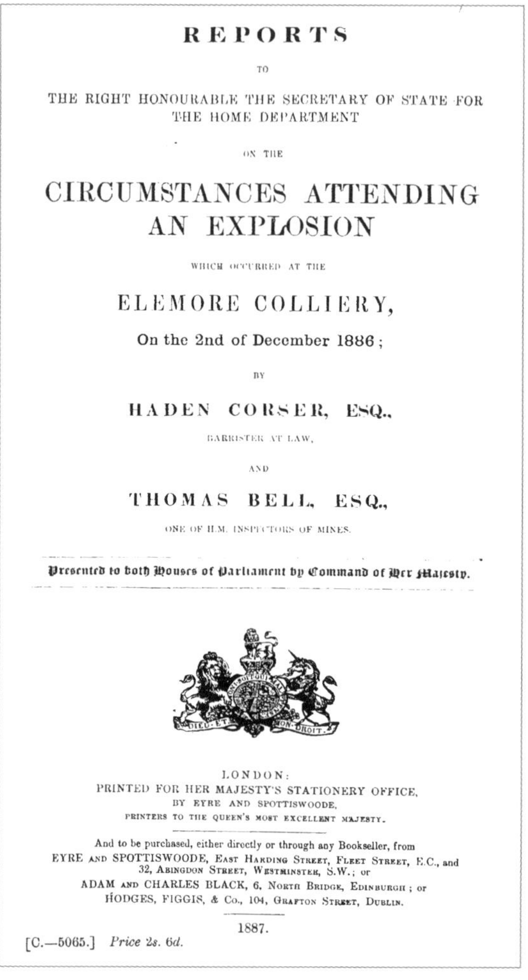

Plate XVIII - Reports ... on the circumstances attending an explosion....(1887).
(Reproduced by permission of HMSO).

The official enquiry into the explosion contained two reports – by Haden Corser, Barrister, and Thomas Bell, Her Majesty's Inspector of Mines. Bell's report can also be found in the *Reports of The Inspectors of Mines to Her Majesty's Secretary of State for the year 1886.* (London: H.M.S.O., 1887).

Officials in charge of the Colliery.

One official was always underground so long as men were down, except at week ends, and when the pits were idle.

Master wasteman - - 2.0 a.m. until 9.0 a.m.
Overmen - - - 3.30 ,, ,, 9.45 ,,
Back overmen - - 9.30 ,, ,, 5.30 p.m.
Assistant master shifter - 5.15 p.m. ,, 1.15 a.m.
Master shifter - - 6.15 ,, ,, 2.15 ,,

The system of working adopted is generally by pillar and stall, with a little long wall working in some of the districts.

The inclination or dip of the seams is towards the east, at an angle of 1 in 50.

Section of the Main Coal Seam in the East or Dale Way :—

	ft.	ins.	
Top coal	- 1	3	
Stone band	-	3	
Gray coal	-	3	ft. ins.
		—— 1	9 not worked.
Good coal worked		3	4
Bottom coal	-		8
		—— 5	9

Section of the George Low Main Coal Seam :—

	ft.	ins.
Good coal	- 3	2
Splint - -		4 ft. ins.
	—— 3	6

Section of the Hutton Seam :—

	ft.	ins.
Good coal	- 4	2
Bottom coarse -		9 ft. ins.
	—— 4	11

Powder was used for blasting coal during the day, but was only allowed to be used by deputies, or other competent men appointed in writing for that purpose. The requirements of the Act were strictly carried out in all respects. All blasting in stone work was done at nights when the ordinary work persons were out of the mine.

The following is a list of the names, ages, and occupations of persons down Elemore Colliery, at the time of the explosion, on December 2, 1886 :—

Seam.	Name.	Age.	Occupation.	Time and Date of Descent.	No.	Remarks.
Lady Hutton Seam.	R. Hills	64	Deputy	8 p.m. Dec. 1st.	1	Killed.
	M. Tempest	38	Hewer	10 ,, ,,	2	,,
	W. Hunter	40	,,	10 ,, ,,	3	,,
	W. Seeds	41	,,	10 ,, ,,	4	,,
	J. Carr	65	,,	10 ,, ,,	5	,,
	G. Nicholson	21	Putter	6 ,, ,,	6	,,
	G. Walton	17	,,	6 ,, ,,	7	,,
	R. Fishburn	60	Horsekeeper	8 ,, ,,	8	,,
George Hutton Seam.	E. Egglestone		Fitter	2.30 a.m. Dec. 2nd.	9	Rescued.
	H. Moss		Horsekeeper	,, ,,	10	,,
Lady Low Main.	H. Johnson		Rapperman	6 p.m. Dec. 1st.	11	,,
	Wm. Johnson		Stoneman	11 ,, ,,	12	,,
	J. Gleghorn		,,	11 ,, ,,	13	,,
	Wm. Johnson		Driver	6 ,, ,,	14	,,
	R. Bousfield		Shifter	6 ,, ,,	15	,,
	G. Gustard		,,	6 ,, ,,	16	,,
	H. Johnson, junr.		Stoneman	11 ,, ,,	17	,,
	H. Buckingham	22	,,	2 a.m. Dec. 2nd.	18	Died since.
	Frank Straughan	32	,,	,, ,,	19	,,
East Main Coal (Dale Way).	T. Hope		Stoneman	11 p.m. Dec. 1st.	20	Rescued.
	Ralph Corner		Putter	6 ,, ,,	21	,,
George Low Main.	Sam Grice	29	Stoneman	2 a.m. Dec. 2nd.	22	Killed.
	Geo. Pattison	54	,,	2 ,, ,,	23	,,
	Wm. Robson	43	Hewer	10 p.m. Dec. 1st.	24	,,
	Jos. Williams	37	,,	10 ,, ,,	25	,,
	S. Parkinson	27	,,	10 ,, ,,	26	,,
	R. Pearson	54	,,	10 ,, ,,	27	,,
	Thos. Robins	20	Putter	6 ,, ,,	28	,,
	Thos. Clark	51	Deputy	8 ,, ,,	29	,,
	J. G. Laverick	22	Stoneman	11 ,, ,,	30	,,
	Thos. Charlton		Master wasteman	2 a.m. Dec. 2nd.	31	Rescued.
	John Johnson	58	Stoneman	11 p.m. Dec. 1st.	32	Killed.
	Geo. Pattison	31	,,	11 ,, ,,	33	,,
	Thos. Johnson		,,	11 ,, ,,	34	Rescued.
	R. Appleby	53	,,	11 ,, ,,	35	Killed.
	John Luke	38	,,	11 ,, ,,	36	Died since.
	G. Thompson	43	Hewer	10 ,, ,,	37	Killed.
	Jno. Thompson	19	,,	10 ,, ,,	38	,,
	[illegible] Taylor	17	Putter	10 ,, ,,	39	,,
	Ralph Lawson	44	Electric engineman	6 ,, ,,	40	,,
Lyons Low Main.	Thos. Spence	36	Furnaceman	8 ,, ,,	41	,,

Plate XIX - Reports of The Inspectors of Mines ... 1886. p. 161.
(Reproduced by permission of HMSO).

Both reports included a list of the dead, and there was also a diagram of the underground seams showing the exact location of the bodies.

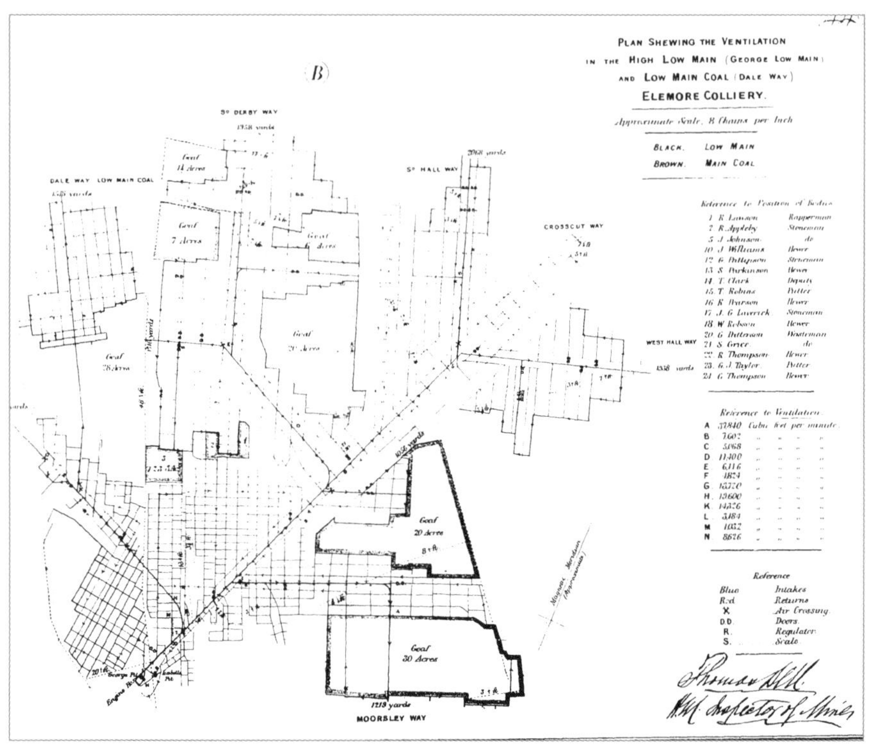

Plate XX - Plan shewing the Ventilation of Elemore Colliery.
(Reproduced by permission of HMSO).

So it is possible to know the exact place, time and circumstances of the death of a coalmining ancestor.

THE EXPLOSION AT ELEMORE COLLIERY.

TELEGRAM FROM THE HOME SECRETARY.

A telegram has been received at Elemore to-day from the Home Secretary (Mr Henry Matthews), directed to the Government Inspector of Mines for the County of Durham. It runs: "Please express condolence with sufferers in accident. Am anxiously awaiting further particulars."

THE INQUEST.

This afternoon Mr Crofton Maynard, coroner for the Easington Ward, held an inquest at the Three Tuns Inn, adjoining the colliery, on the bodies of four of the five men recovered from the pit. Their names are:—Ralph Fishburn, Thomas Spence, Ralph Appleby, and Ralph Lawson. Mr Chapman, of Easington Lane, was chosen foreman of the jury. Amongst those present were—Sir H. Havelock-Allan, Bart., M.P., Mr W. Crawford, M.P. (secretary to the Durham Miners' Union), Mr W. H. Patterson (financial secretary), Mr J. W. Willis (Government Inspector of Mines), Dr. Adamson, and Mr T. Wood, representing the colliery owners.—The Coroner, addressing Mr Crawford, said: Can you give me any idea when the rest of the bodies will be brought out?—Mr Crawford: Not the slightest.—The Coroner said he only made the inquiry in order that they might get the bodies identified.—Mr Patterson said he believed one body would be got out to-night.—The jury having been sworn proceeded to the mortuary on the pit bank and viewed the bodies of the four deceased.—Formal evidence of identification was then taken.—The inquiry had not terminated when we went to press.

Plate XXI - Sunderland Daily Echo. 3 December 1886. p. 4.

The explosion was of national importance, as can be seen from this newspaper report of the receipt of a telegram from the Home Secretary. The inquest opened at a public house near the colliery on 20 December 1886.

After the funerals, a memorial was erected in front of the church, which stands just off the High Street in Easington Lane. It lists the names of all those who lost their lives in 1886.

CHAPTER 11
National Collections

Coalmining was largely a provincial industry, so in the search for coalmining ancestors it is not always necessary to visit the national collections to use their over-stretched resources.

Archives

The National Archives, Kew

The single most important repository of documentary information on individuals in Britain. The vast range of government enquiries and reports into the coal industry has been mentioned in Chapter 2, and there is much documentary evidence of this in TNA. Accident reports, health and safety, inspection, industrial disputes, nationalisation, miners' welfare and wartime measures are all part of the vast holdings. The leaflets *Coal Mining Records in the National Archives* (Domestic Records Information 35, last updated 2007) and *Sources for the History of Mines and Quarries* (Domestic Records Information 64, last updated 2007) are indispensable guides to these records, and can be downloaded from the website. They contain links to the various classes of records in the archive. It should be noted, however, that TNA does not hold mining personnel records, which are more likely to be held in the former coalfield areas (see Chapter 12).

 www.nationalarchives.gov.uk/

The National Archives of Scotland, Edinburgh

The National Archives of Scotland holds the main historical records of coal mining in Scotland – NCB records as well as those of the earlier coal companies before nationalisation. There is an online guide to coalmining records.

 www.nas.gov.uk/guides/coalmining.asp

The Coal Authority

Although some material from the former National Coal Board/British Coal Corporation was transferred to the Coal Authority, little of it is available for consultation, or is of immediate value to the family historian. Personnel records are not held, although public access to plans of abandoned mines is available. A brochure on the Mining Records and Reports Service is available for download from the website.

 www.coal.gov.uk/media/E6A8A/mrsds_brochure_2007.pdf

Iron Mountain

On behalf of the government, Iron Mountain manages over 3.6 million employment records of all employees of the former British Coal Corporation. These are stored in a former coalmine at Cannock in Staffordshire. Two 'case studies' on the service can be downloaded from the website.

 www.ironmountain.co.uk/resource/casestudies/DepartmentofTrade andIndustry.pdf

 www.ironmountain.co.uk/resource/casestudies/berr-case-study.pdf

Libraries

The British Library

In its new home at St Pancras, the British Library has the largest collection of printed books in the country, together with manuscript collections, maps and photographs. A Reader's Pass is required for the Reading Rooms, although the catalogues of the library, printed and on-line, are invaluable in tracking-down the existence and bibliographical details of printed material. The **British Library Newspaper Library** at Colindale has less restricted access, and includes all national and local, daily and weekly newspapers, and periodicals, of the last two hundred years.

 www.bl.uk/

 www.bl.uk/reshelp/inrrooms/blnewspapers/newsrr.html

The National Library of Scotland

The Manuscript Division of the Department of Special Collections houses a number of records relating to coalmining. There are papers relating to various Scottish mine-owning families; and trade union records of the National Union of Mineworkers, Scotland, and its branches, including the original drafts of R. Page Arnot's *History of the Scottish Miners* (1955). The collection also includes the correspondence and papers of Willie Hamilton, former M.P. for Fife and a Durham miner's son.

The 'Family history – frequently asked questions' section of the website provides useful general advice.

 www.nls.uk/family-history/questions.html

The National Library of Wales

The National Library of Wales is acknowledged as the principal centre for researching Welsh genealogy, and welcomes family historians to use its sources and facilities. Its website includes a section on 'Family History'. There is an online catalogue to its collections, although a reader's ticket is required to access the Reading Rooms.

 http://www.llgc.org.uk/index.php?id=familyhistory0

Museums

England, Scotland and Wales each have a national museum of coalmining. None of them would claim to have material of direct relevance in the search for coalmining ancestors, although they all have evocative displays and information that help to explain the life and work of a coalminer. They also have well-stocked libraries to provide further information for the researcher. Over the years they have also produced leaflets and booklets providing guidance for the family historian

National Coal Mining Museum for England

Based at the former Caphouse Colliery halfway between Wakefield and Huddersfield West Yorkshire. It includes the oldest mineshaft still in everyday use.

An underground tour is available, as well as exhibitions and collections on the themes of mining art, oral history, engineering and technology, social history and a photographic archive. It is possible to view many of the items online. The library holds transcripts and summaries of the oral history recordings, as well as books, reports and journals. There is also a picnic area, gift shop, and restaurant. More information can be found on page 112.

 www.ncm.org.uk/

Scottish Mining Museum

On the site of the Lady Victoria Colliery at Newtongrange, nine miles south of Edinburgh. The Visitor Centre contains two major exhibitions – 'The Story of Coal' and 'A Race Apart'. There is a library and archive which can be visited by appointment, and a large photographic collection. The museum has published a number of 'Fact files' for adults and children which provide useful background information. See pages 6, 63 and 145,

 www.scottishminingmuseum.com/

Big Pit Mining Museum

This museum is at Blaenafon near Newport in South Wales, and is the national mining museum for Wales. There is an underground tour, and surface buildings which now house the galleries and exhibitions. There is a Resource Room for researchers, as well as a canteen and gift shop. Go to page 137 for more information.

 www.museumwales.ac.uk/en/bigpit/

 www.museumwales.ac.uk/en/rhagor/category/?cat=75

The People's History Museum

Formerly also known as the National Museum of Labour History, and based in Manchester, this museum portrays the strong links between miners and the labour and trade union movement. There are significant collections of political cartoons and banners together with an archive and study centre which includes the archives of the Labour Party, the Independent Labour Party and the Communist Party. A new building opened in February 2010. See page 107.

 www.phm.org.uk

Societies and Institutions

Society of Genealogists

Although a private institution, this society has a worldwide membership. It owns the largest collection of genealogical books and material in the country – parish registers and bishops' transcripts, monumental inscriptions and censuses, local history, as well as an unequalled assembly of research notes on families. It publishes the quarterly *Genealogists' Magazine* and a number of books (including this one) and information leaflets. In the coming years the society plans to make as much of its collection as possible available over the Internet. It also runs a regular programme of lectures, talks and tutorials throughout the year.

 http://www.sog.org.uk/

Although all of the collections mentioned above contain unique items, which can only be inspected on a personal visit, technology is allowing more and more material to be microfilmed, digitised, indexed, or copied and made available at local centres (see Chapter 12), or over the Internet (see Chapter 13).

Details of national collections which are located outside London may be found in the appropriate section on each coalfield in Chapter 12.

CHAPTER 12
Coalfield collections

Within the coalfields of England, Scotland and Wales almost every archive, family history society, library, museum or record office will have material of potential value to anyone with a coalmining ancestor. However, before making a special journey to visit it is always desirable to have some prior knowledge of what is available. More and more libraries and record offices are making their catalogues available over the Internet, to enable some preliminary searching to take place. There may also be down-loadable images and information leaflets.

 http://dspace.dial.pipex.com/town/square/ac940/weblibs.html

 www.archiveshub.ac.uk/

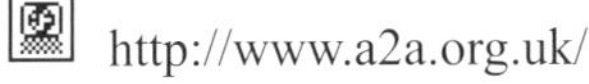 http://www.a2a.org.uk/

This chapter identifies those with collections of particular interest or importance. Many of these organisations also publish material of interest, and some of these items are listed below.

ENGLAND

Northumberland and Durham

The coalfield is about 60 miles north to south and 800 square miles in an area roughly the shape of a triangle, the apex of which is near the mouth of the River Coquet in Northumberland and the base on a line from Middleton in Teesdale to Hartlepool. It lies in the historic counties of Northumberland and Durham (now Northumberland, Tyne and Wear, and Durham) in the North East of England. This was the oldest coalfield in the country, coal being first mined here in Roman times.

 *Durham County Record Office*

User and subject guides can be downloaded from the website, and the catalogue (including that of the library) is searchable on-line.

 N.C.B. deposits; Durham Miners Association records; Durham coalowners Association records; Durham Colliery Owners' Mutual Protection Association records; NACODS (Durham Area) records; Durham County Colliery Enginemen's, Boiler Minders' and Firemens' Association records; Mining Records Project records

County Hall, Durham, DH1 5UL

0191 383 3253

record.office@durham.gov.uk

www.durham.gov.uk/recordoffice
There is a link from the Home Page to a searchable database of information on Durham collieries.

 *Subject Guide 7: Colliery Personnel Records*
A list which provides brief details of the records of collieries which contain information on individual miners.

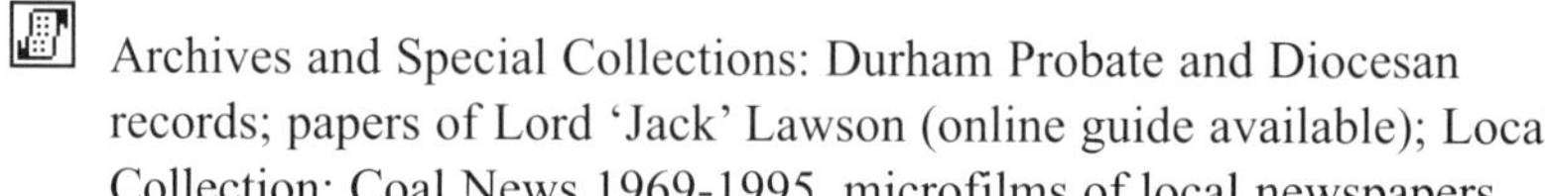 *Durham University Library*

Archives and Special Collections: Durham Probate and Diocesan records; papers of Lord 'Jack' Lawson (online guide available); Local Collection: Coal News 1969-1995, microfilms of local newspapers.

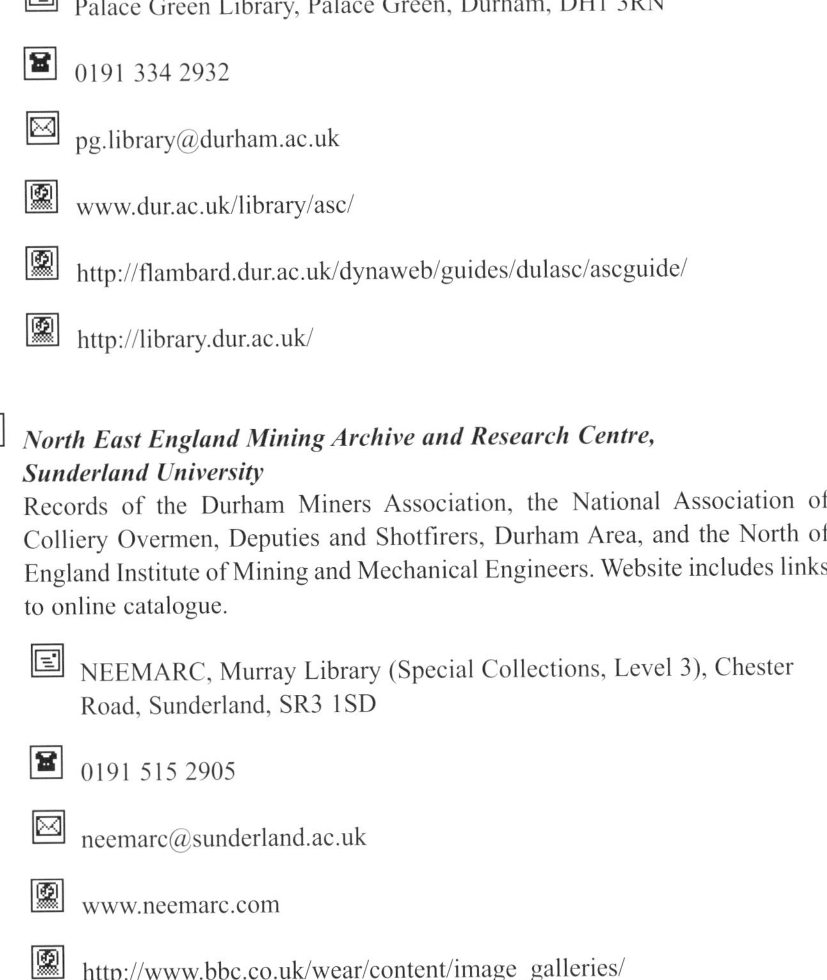

Palace Green Library, Palace Green, Durham, DH1 3RN

0191 334 2932

pg.library@durham.ac.uk

www.dur.ac.uk/library/asc/

http://flambard.dur.ac.uk/dynaweb/guides/dulasc/ascguide/

http://library.dur.ac.uk/

North East England Mining Archive and Research Centre, Sunderland University

Records of the Durham Miners Association, the National Association of Colliery Overmen, Deputies and Shotfirers, Durham Area, and the North of England Institute of Mining and Mechanical Engineers. Website includes links to online catalogue.

NEEMARC, Murray Library (Special Collections, Level 3), Chester Road, Sunderland, SR3 1SD

0191 515 2905

neemarc@sunderland.ac.uk

www.neemarc.com

http://www.bbc.co.uk/wear/content/image_galleries/mining_archive_gallery.shtml

Tyne and Wear Archives

Includes records relating to the whole coalfield. Online catalogue available. User Guides can be downloaded from the website.

Bonds; Viewers' reports; papers re the Hartley Colliery Disaster, 1862

Blandford House, Blandford Square, Newcastle upon Tyne, NE1 4JA

📞 0191 232 6789

✉ TWAS@gateshead.gov.uk

🌐 www.tyneandweararchives.org.uk

ℹ️ User Guide 19A Coal Industry. (2006). Provides details of records relevant to a study of the history of the industry in the North East, with records from 1710. Material relating to the Northumberland and Durham Miners Permanent Relief Fund is especially useful for the family historian.

ℹ️ User Guide 19B: Collieries. (2006)
A list of the records relating to individual collieries.

🅰 *Special Collections, Robinson Library, Newcastle University*
Online catalogue available

Sid Chaplin and Jack Common papers; 17th-19th century coal trade

Newcastle upon Tyne, NE2 4HQ

📞 0191 222 7671

🌐 www.ncl.ac.uk/library/specialcollections

🅰 *Woodhorn*

A new facility incorporating the former County Record Office, County Library Local Studies Collection; 3-D collections relating to coalmining (now known collectively as Northumberland Collections Service); and Woodhorn Colliery Museum. Online catalogue available.

N.C.B. deposits; Northumberland Coal Owners Mutual Protection Association; Northumberland Colliery Mechanics Association; N.U.M. branch records; paybills; Bonds; database of photographs Individual coal company and mine-owners records.

Coal Town (coalmining in Northumberland); Banner Ramp (miners' banners); The Ashington Group (pitmens' paintings); temporary exhibitions; mining artifacts.

Woodhorn, Northumberland Museum and Archives, Queen Elizabeth II Country Park, Ashington, Northumberland, NE63 9YF

01670 528080

www.experiencewoodhorn.org.uk

www.northumberland.gov.uk/collections

collections@woodhorn.org.uk

Northumberland Family History-a brief guide (2006) (downloadable from website)

Durham Clayport Library

Online catalogue available

'The Durham Record'– an interactive online database of photographic images and digitised Ordnance Survey maps of 1860, 1898, 1923, 1996; 'The Durham Miner' – project now ceased but information still available online

Durham Clayport Library, Millennium Place, Durham, DH1 1WA

0191 386 4003

durhamclayportlibrary@durham.gov.uk

www.durham.gov.uk/

www.durham-miner.org.uk

Gateshead Library

Online catalogue available

Oral history – Reminiscences of the 1921 and 1926 strikes; newspapers relating to the General Strike 1926

Central Library, Prince Consort Road, Gateshead, NE8 4LN

0191 433 8400

www.gateshead.gov.uk/ls

Bennett, G *et al, A fighting trade: rail transport in Tyne coal
1600-1800. 2 vols.* (Gateshead: Portcullis Press, 1989).

Newcastle Library

Local Studies Centre covers the whole region; the Humanities Library also includes
material of interest. New Central Library opened in 2009. Online catalogue available.

Diary of Matthias Dunn 1831-36; Reports of H.M. Inspectors of Mines
1859- ; Picture Collection.

Local Studies Centre, Exhibition Hall, Civic Centre, Newcastle upon
Tyne, NE1 8PU

0845 002 0336

information@newcastle.gov.uk

www.newcastle.gov.uk

North of England Institute of Mining and Mechanical Engineers

Online catalogue of books and journals available via the website

Neville Hall, Westgate Road, Newcastle upon Tyne, NE1 1SE

0191 232 2201

librarian@mininginstitute.org.uk

www.mininginstitute.org.uk

Sunderland Library

Online catalogue available

Transactions of the North of England Institute of Mining Engineers
1852-1916

Local Studies Centre, City Library and Arts Centre, Fawcett Street, Sunderland, SR1 1RE

0191 514 8439

local.studies@sunderland.gov.uk

www.sunderland.gov.uk

 *Beamish – The North of England Open Air Museum*

Regional Resource Centre includes Photographic archive, Reference library, Audio history archive: Miners' Lodge Banners

Colliery Village 1913 (open during Summer season only), including Pit cottages and gardens, Chapel, Board school, Drift Mine, Steam winder, Pithead buildings

Beamish – The North of England Open Air Museum, Beamish, Stanley, Co. Durham, DH9 0RG

0191 370 4000

museum@beamish.org.uk

www.beamish.org.uk

www.beamishcollections.com

downloadable Visitor Guide

Doyle, A, *The Great Northern Coalfield: mining collections at Beamish Museum* (Newcastle upon Tyne: Northumbria University Press, 2005).

Durham Mining Museum

Originally a 'virtual' museum, with an ever-increasing amount of online information; now with a physical presence (on Tuesdays and Thursdays) in Thornley.

Memorial Roll of names of those killed in mining-related accidents –
also available online

Miners' banners; mining equipment

Thornley Community Centre, Hartlepool Street, Thornley, Co. Durham,
DH6 3AB

www.dmm.org.uk

F Pit Museum, Washington

The only surviving winding engine house in situ in the Durham coalfield, but
now surrounded by housing development in a new town. The colliery opened
in 1777 and closed in 1968. It opened as a museum in 1976, and is now owned
by Sunderland City Council and managed by Tyne and Wear Museums.
Limited opening hours during summer months only. Contact Sunderland
Museum for further information.

www.bbc.co.uk/wear/content/articles/
2007/11/20/fpit_washington_feature.shtml

Sunderland Museum

After a history of 150 years this museum has been totally revamped.

'Coal': a lasting tribute to nearly 200 years of mining in the East
Durham coalfield

Sunderland Museum and Winter Gardens, Burdon Road,
Sunderland, SR1 1PP

0191 553 2323

www.twmuseums.org.uk/sunderland/

Useful publications:

General description of coalfield

Atkinson, F *Great Northern Coalfield 1700-1900*. (London: University
Tutorial Press, 1968). An illustrated history of one of the oldest coalfields.

Hair, T H *A series of views of the collieries in the counties of Northumberland and Durham; with descriptive sketches and a preliminary essay on coal and the coal trade, by M. Ross*. (Newcastle upon Tyne: Davis Books, 1987), originally published 1844. Classic descriptions and illustrations.

Kirkup, M ed. *Eyewitness: the Great Northern Coalfield*. (Newcastle upon Tyne: TUPS Books, 1999). Consists in the main of reprinted contemporary documents and descriptions of the Northumberland and Durham coalfield from 1725 to 1985. Includes some fascinating illustrations, some from Hair (see above), but there is no index and few individual miners are named.

Page, W, ed. *Victoria History of the County of Durham*, ed. W. Page, vol. 2. (London: University of London, Institute of Historical Research, 1968). pp. 175-258; 320-348. Includes a chronological list of mining development, pp. 329-345.

Mining histories

Colls, R *Pitmen of the Northern coalfield: work, culture and protest, 1790-1850*. (Manchester: Manchester University Press, 1987). The history of the Great Northern Coalfield during the crucial formative years of change.

Durham Aged Mineworkers' Homes Association *Annual Reports 1898-1998*. Scanned, searchable images of the Annual Reports, accounts, booklets and other records of the Association on CD-ROM.

Emery, N *Coalminers of Durham*. (Stroud: Sutton Publishing, 1992). A lavishly illustrated history.

Fynes, R *Miners of Northumberland and Durham: a history of their social and political progress*. (Newcastle upon Tyne: Davis Books, 1986, first published 1873). The classic first-hand account.

Garside, W R *Durham miners 1919-1960*. (London: Allen & Unwin, 1971). The official history of the Durham miners, including a bibliography and index. Few individual miners are mentioned by name, but the book does include an anonymous description of 'Life in a Durham Pit Village in the 1920s'.

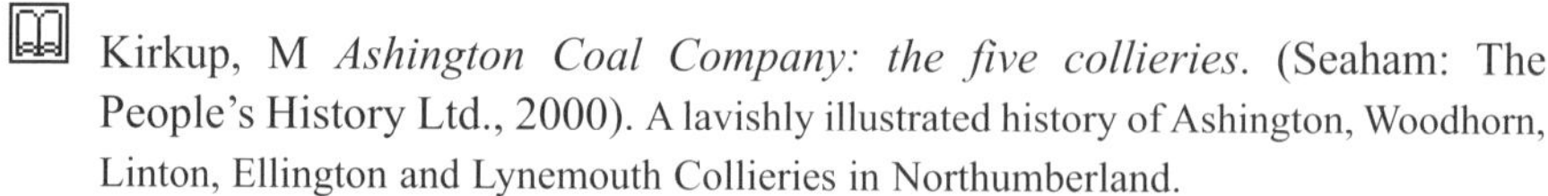

Kirkup, M *Ashington Coal Company: the five collieries.* (Seaham: The People's History Ltd., 2000). A lavishly illustrated history of Ashington, Woodhorn, Linton, Ellington and Lynemouth Collieries in Northumberland.

Smith, K and Smith J *The Great Northern miners.* (Newcastle upon Tyne: Tyne Bridge Publishing, 2008). An extensively illustrated history of the miners of the two counties.

Temple, D *Collieries of Durham.* 2 vols. (Newcastle upon Tyne: TUPS Books, 1994 and 1995). A history of the coalfield accompanied by many evocative illustrations.

Tuck, J T *Collieries of Northumberland.* 2 vols. (Newcastle upon Tyne: TUPS Books, 1993 and 1995). Descriptions and photographs of 38 collieries. Volume 2 includes a brief illustrated history of miners' unions in Northumberland and Durham.

Social history

Griffiths, B *Pitmatic: the talk of the North East coalfield* (Newcastle upon Tyne: Northumbria University Press, 2007).

Moore, R S *Pit-men, preachers and politics: the effects of Methodism in a Durham mining community.* (Cambridge: Cambridge University Press, 1974). A now-classic study of the decline of the influence of Methodism on the mining community, and of the community itself, in the late nineteenth and early twentieth centuries.

Thompson, R *How long did the ponies live?: the story of the colliery at Killingworth and West Moor.* (North Shields: Beacon House, 1997). A chronological history of a famous mine and the daily life of its people.

Union histories

Davison, J *Northumberland miners 1919-1939.* (Newcastle upon Tyne: NUM (Northumberland Area), 1973). Includes the political background to the history of the N.U.M. in Northumberland. Contains some photographs and is indexed.

Hall, W S *Historical survey of the Durham Colliery Mechanics' Association 1879-1929*. (Durham: J.H. Veitch, 1929). A volume which is difficult to use as there is no index. However, if your ancestor was an official of the association you could well find his photograph.

Wilson, J *History of the Durham Miners' Association 1870-1904*. (Durham: J.H. Veitch, 1907). The standard history of the DMA by a prominent miners' representative. Includes photos of officials. Now available on CD ROM.

Biographies/autobiographies

Lawson, J *Peter Lee*. (London: Hodder, 1936). The story of how Peter Lee worked, wandered, fought, and finally led miners and communities, by a M.P. who knew and admired him whole heartedly.

Websites

www.bbc.co.uk/nationonfilm/topics/coal-mining/ *Nation on Film*. Evocative film clips of coalmining life in the North East in the twentieth century.

www.bbc.co.uk/wear/history/mining/ *Mining in the North East.*

Cumberland

The Cumberland coalfield was about 25 miles long and 6 miles wide along the Irish Sea coast from Barrowmouth near Whitehaven to Maryport. It had an area of about 108 square miles.

The Beacon, Whitehaven

A recently redeveloped award-winning museum, gallery and archive overlooking the harbour which tells the story of Copeland's rich history.

Showcase of mining artifacts including commemorative items and other equipment

West Strand, Whitehaven, Cumbria, CA28 7LY

01946 592302

thebeacon@copelandbc.gov.uk

 www.thebeacon-whitehaven.co.uk

Haig Colliery Mining Museum

Library catalogue (MS Access) can be downloaded.

Oral history tapes

Solway Road, Kells, Whitehaven, Cumbria, CA28 9BG

01946 599949

museum@haigpit.com

www.haigpit.com
Includes description of mining in Whitehaven in 1814

Information Pack available

Useful publications:

General description of coalfield

Wilson, J ed. *Victoria History of the County of Cumberland, vol. 2*, (London: Constable, 1905). pp. 348-384. Includes a list of the 43 collieries which were operating in the year 1900, and detailed listings, by parish, of mining developments.

Wood, O *West Cumberland Coal 1600-1982/3*. (Kendal: Cumberland and Westmorland Antiquarian and Archaeological Society, 1988). Detailed history of this little-known coalfield. Includes a glossary and an extensive bibliography.

Mining histories

Beckett, J V *Coal and tobacco: the Lowthers and the economic redevelopment of West Cumberland 1660-1760*. (Cambridge: Cambridge University Press, 1981). One hundred years in the history and development of the coalfield and the prominent part played therein by Sir John Lowther and his son Sir James, later Earl of Lonsdale.

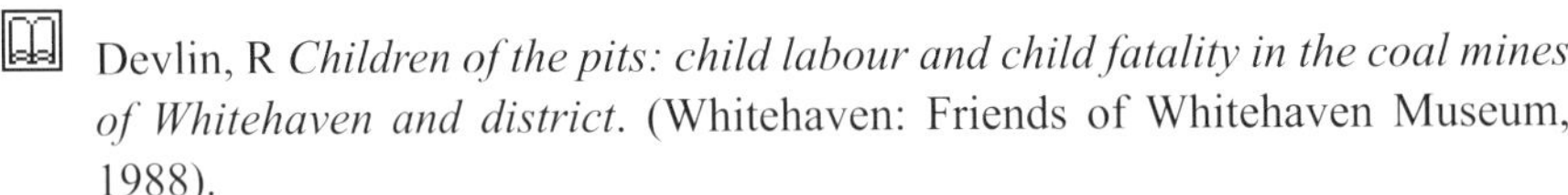

Devlin, R *Children of the pits: child labour and child fatality in the coal mines of Whitehaven and district.* (Whitehaven: Friends of Whitehaven Museum, 1988).

Devlin, R and Fancy, H *The most dangerous pit in the kingdom.* (Workington: Hills Books, 1997).

Fisher, M and Donnelly, S *"Ah'd gaa back tomorra!": memories of West Cumbrian screen lasses* (Whitehaven: Whitehaven Miners' Memorial and Living History Project, 2004).

Lancashire & Cheshire

The Lancashire and Cheshire field extended some 500 square miles to the north of Manchester, with only a very small part in Cheshire. Somewhat romantically, Sir Arthur Trueman described its shape as 'an axe with the blade pointed to the west and the handle projecting southwards'.

Bolton Museum and Archive Service

Online catalogue available

Hulton Colliery (Pretoria Pit) Disaster Relief Fund; Ladyshore Coal Company records

Museum and Archive Service, Le Mans Crescent, Bolton, BL1 1SE

01204 332185

archives.library@bolton.gov.uk

www.boltonmuseums.org.uk

Cheshire and Chester Archives and Local Studies Service

Online catalogue available

Records of Oakmere Rehabilitation Centre 1944-1974; Vernon collection red Poynton and Worth collieries; Shakerley Coal Company

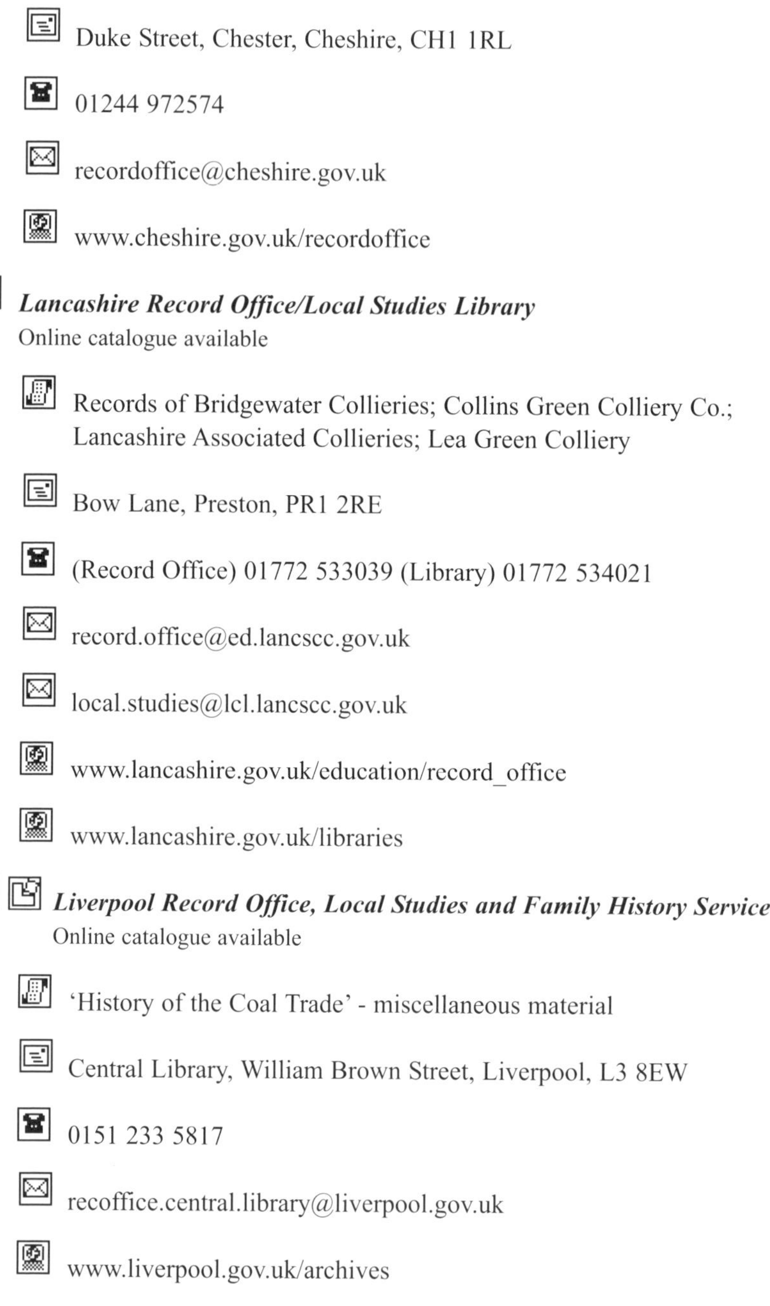

Duke Street, Chester, Cheshire, CH1 1RL

01244 972574

recordoffice@cheshire.gov.uk

www.cheshire.gov.uk/recordoffice

Lancashire Record Office/Local Studies Library
Online catalogue available

Records of Bridgewater Collieries; Collins Green Colliery Co.; Lancashire Associated Collieries; Lea Green Colliery

Bow Lane, Preston, PR1 2RE

(Record Office) 01772 533039 (Library) 01772 534021

record.office@ed.lancscc.gov.uk

local.studies@lcl.lancscc.gov.uk

www.lancashire.gov.uk/education/record_office

www.lancashire.gov.uk/libraries

Liverpool Record Office, Local Studies and Family History Service
Online catalogue available

'History of the Coal Trade' - miscellaneous material

Central Library, William Brown Street, Liverpool, L3 8EW

0151 233 5817

recoffice.central.library@liverpool.gov.uk

www.liverpool.gov.uk/archives

St. Helens Local History and Archives Library

Mining deaths in Great Britain (7 volumes); Index of mining accidents

Central Library, Gamble Institute, Victoria Square, St. Helens, Merseyside, WA10 1DY

01744 456952

localhistory&archivesservices@sthelens.gov.uk

www.sthelenshistory.net

Salford Local History Library and City Archives Centre
Online catalogue available

Newtown Colliery Welfare & Benevolent Fund records; Pendlebury Branch, N.U.M. records including list of miners employed

Salford Museum and Art Gallery, Peel Park, The Crescent, Salford, M5 4WU

0161 778 0814

local.history@salford.gov.uk

www.salford.gov.uk

Working Class Movement Library
Collection of books and other media on the labour movement. Online catalogue available.

Jubilee House, 51 The Crescent, Salford, M5 4WX

0161 736 3601

enquiries@wcml.org.uk

www.wcml.org.uk

Westhoughton Library, Bolton

Online catalogue available

Extensive collection on the Hulton Colliery (Pretoria Pit) Disaster of 21 December 1910

Westhoughton Library, Library Street, Westhoughton, Bolton, Lancashire, BL5 3AU

www.bolton.gov.uk/libraries

Catalogue of material. 4th ed. (1993).

Astley Green Colliery Museum

The colliery opened in 1908 and closed in 1970. Now fully restored as a working museum by the Red Rose Steam Society. Open Sunday, Tuesday and Thursday afternoons, 1.00 p.m.-5.00 p.m.

Extensive displays of industrial mining and related artifacts

Higher Green Lane, Astley, Tyldesley, Manchester, M29 7JB

01942 828121

info@agcm.org.uk

www.agcm.org.uk

The History Shop, Wigan

"Founded on coal" in the Taylor Gallery

The History Shop, Library Street, Wigan,

01942 828128

heritage@wlct.org

www.wlct.org/heritage

Museum of Local Crafts and Industries, Burnley

Items of mining equipment; photographs; documents regarding wages and disasters

Towneley Hall Museum, Towneley Park, Burnley, BB11 3RQ

01282 424213

towneleyhall@burnley.gov.uk

www.burnley.gov.uk/towneley/site/index.php

People's History Museum - Archive and Study Centre
The museum reopened in February 2010.

Labour Party archives

Left Bank, Spinningfields, Manchester, M3 3ER

0161 838 9190

archives@phm.org.uk

www.phm.org.uk

Useful publications:

General description of coalfield

Farrer, W and Brownbill, J eds. *Victoria History of the County of Lancaster*, vol. 2. (London: Constable, 1908). pp. 355-359.

Mining histories

Anderson, D *Orrell coalfield, Lancashire 1740-1850*. (Buxton: Moorland Publishing, 1975).

Davies, A and Hudson, L *The Wigan Coalfield*. (Images of England series). (Stroud: Tempus Publishing, 1999). Contains many photographs of the coalfield and tells the long history of the town's association with coal.

Nadin, J *East Lancashire mining memories*. (Stroud: The History Press, 2008).

Social history

Forman, C *Industrial town: self portrait of St Helens in the 1920s*. (London: Granada, 1979). A unique collection of eyewitness accounts of life in a mining town in the aftermath of the Great War.

Naylor, F H *Lord Crawford's other acre*. (Wigan: F.H. Naylor, 1993). The history of a one-street Lancastrian mining village, where the colliery closed in October 1940, forms the background of a history of the Naylor and Hughes families.

Simm, G and Winstanley, I *Mining Memories*. (St. Helens: St. Helens M.B.C. Community Leisure Dept., 1990). Contains over 150 illustrations and copies of documents relating to the history of coal mining in St Helens.

Union histories

Challinor, R *Lancashire and Cheshire miners*. (Newcastle upon Tyne: F. Graham, 1972). The history of trade unionism in the area from its beginnings to the end of the nineteenth century. Contains detailed notes and bibliography, and is indexed.

Howell, D *The politics of the N.U.M.: a Lancashire view*. (Manchester: Manchester University Press, 1989). An analysis of NUM politics based on a study of the Lancashire area during the years up to and including the strike of 1985.

Biographies/autobiographies

Howarth, H *Dark days: memories of the Lancashire and Cheshire coalmining industry up to nationalisation*. (Manchester: Greater Manchester County Council, 1978, reprinted with new introduction, 1985). Transcriptions of the reminiscences of former miners compiled after an oral history survey.

 www.jnadin1.50megs.com/ *Jack Nadin's Lancashire Coal Mining History Page*

Yorkshire

The largest and most important coalfield in the United Kingdom, it extended from Leeds to Nottingham, about 60 miles, with some 1,400 square miles of proven coalfield in the former county of the West Riding of Yorkshire; and continued into Derbyshire and Nottinghamshire (see next section).

 *John Goodchild Collection*

A private collection of manuscripts and books relating to the West Riding of Yorkshire, open free of charge to all, by appointment.

 Material on coalmining relates to owners, officials and trade union leaders only

 John Goodchild, M.Univ., Local History Study Centre, below Central Library, Drury Lane, Wakefield, WF1 2DT

 01924 288929

 www.wakefield.gov.uk

 *Hull History Centre*

A brand new centre for city and university archives and local history
Online catalogue available

 papers of W.E. Jones, President of the NUM; papers of R. Page Arnot, mining historian

 Worship Street, Hull, HU2 8BG

 01482 317500

www.hullhistorycentre.org.uk

Central Library, Shambles Street, Barnsley, S70 2JF

01226 773950

archives@barnsley.gov.uk

www.barnsley.gov.uk

Doncaster Local Studies Library

'The Miners' Strike 1984-85'; local newspapers

Central Library, Waterdale, Doncaster, South Yorkshire, DN1 3JE

01302 734307

central.localhistory@doncaster.gov.uk

www.doncaster.gov.uk/localstudies

www.doncaster.gov.uk/Leisure_in_Doncaster/Libraries/
Archives_Local_Studies/family_local_history/Coal_Industry.asp
Coal industry records in Doncaster Archives

Wakefield Learning and Local Studies Library
Online catalogue available

Mining Deaths in Great Britain 1850-1914

Library Headquarters, Balne Lane, Wakefield, WF2 0DQ

01924 302230

lib.learningandlocalstudies@wakefield.gov.uk

www.wakefield.gov.uk/libraries

Elsecar Heritage Centre

Newcomen Beam Engine, 1795; Y.M.A. banner, Elescar Hemmingfield Colliery

Wath Road, Elescar, Barnsley, S74 8HJ

01226 740203

elsecarheritagecentre@barnsley.gov.uk

barnsley.gov.uk

Museum of South Yorkshire Life

Commemorative items; tools; photographs; ephemera

Cusworth Hall Museum and Park, Cusworth Lane, Doncaster, DN5 7TU

01302 782342

museum@doncaster.gov.uk

www.doncaster.gov.uk/cusworthhall

National Coal Mining Museum for England
The Library has a wide range of books, reports, journals and oral history material. Online catalogue available.

Collections: Art; Engineering and Technology; Oral History; Photographic; Social History

exhibitions and temporary displays illustrating coalmining; nature trail

Caphouse Colliery, New Road, Overton, Wakefield, WF4 4RH

01924 848806

info@ncm.org.uk

www.ncm.org.uk

Henesey, A *Routes to your roots: mapping your mining past* (Overton: National Coal Mining Museum for England, 2004).

Salway, G *The British Coal Collection.* (Overton: National Coal Mining Museum for England, 1999).

Visitor guide; *Tracing your mining family* (downloadable from website).

Wakefield Museum

'Story of Wakefield' Gallery has many coalmining items including a reconstructed miner's kitchen; miners' lamps; strike leaflets, ephemera and photographs

Wood Street, Wakefield, West Yorkshire, WF1 2EW

01924 305356

www.wakefieldmuseums.org

Useful publications:

General description of coalfield

Addy, J *Coal and iron community in the Industrial Revolution*, 1760-1860. (Harlow: Longman, 1969). A book written for children and based on the history of the South Yorkshire coalfield.

Benson, J and Neville, R G eds. *Studies in the Yorkshire coal industry.* (Manchester: Manchester University Press, 1976). A number of individual contributions/topics make up this description of the Yorkshire coalfield.

Mining histories

Booth, A J *Railway history of Denaby Main and Cadeby Collieries.* (Industrial Railway Soc.) Not just a railway book – it also names the people killed in the 1912 explosion and the history of the colliery and many pictures.

Elliott, B *Pits & pitmen of Barnsley*. (Barnsley: Wharncliffe Books).
A pictorial celebration of the men who worked in the pits and the society in which they lived, from Victorian times to the 1990's.

Gill, M C and Davison, P *Keighley coal: a history of coal mining in the Keighley district* (British mining, 74). (Sheffield: Northern Mine Research Society, 2004).

Goodchild, J *Before the National Coal Mining Museum: a new history of Caphouse Colliery and Denby Grange Collieries*. (Wakefield: Wakefield Historical Publications, 2000).

Goodchild, J *The West Yorkshire Coalfield*. (Stroud: Tempus Publishing, 2000).

Social histories

Brooke, A *Colliers and hurriers: working conditions in coalmines of the Huddersfield area c. 1800-1870*. (Workers History Publications, 1992).

Dennis, N *et al, Coal is our life: an analysis of a Yorkshire mining community*. 2nd ed. (London: Tavistock, 1969). A sociological study of 'Ashton', a small town at the heart of the Yorkshire coalfield, in the 1950s, first published in 1956. It provides a vivid description of a way of life as distant from us as is the original colliery of 1868.

Hill, A *The South Yorkshire coalfield: a history and development* (Stroud: Tempus, 2001).

Warwick, D and Littlejohn, G *Coal capital and culture: a sociological analysis of mining communities in West Yorkshire*. (London: Routledge, 1992). An attempt to replicate the classic study by Dennis *et.al*. Based in West Yorkshire during the 1980s, including the period of the miners' strike of 1984/5.

Williams, P *Images of Yorkshire coal* (Ashbourne: Landmark, 2005). A selection of surface and aerial photographs from the collection of the Royal Commission on the Historical Monuments of England. Each photo an explanation and more importantly a grid reference.

Union histories

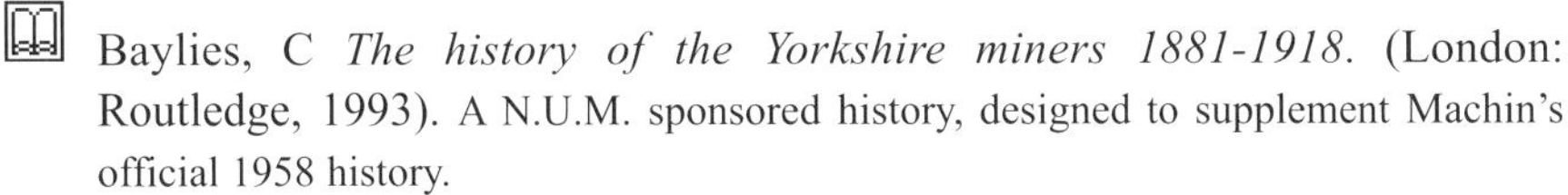

Baylies, C *The history of the Yorkshire miners 1881-1918.* (London: Routledge, 1993). A N.U.M. sponsored history, designed to supplement Machin's official 1958 history.

Machin, F *Yorkshire miners.* vol.1. (Barnsley: NUM (Yorkshire Area), 1958). The official history, illustrated and with an index.

Winterton, J and R *Coal, crisis and conflict: the 1984-85 miners' strike in Yorkshire.* (Manchester: Manchester University Press, 1989). Includes a useful 'Strike Chronology', pp. 295-302.

Biographies/autobiographies

Bailey C *Black diamonds: the rise and fall of an English dynasty.* (London: Viking, 2007). The story of Wentworth House, the family who lived there, and its great estate at the centre of the South Yorkshire coalfield.

Bullock, J *Bowers Row.* (East Ardsley: EP Publishing, 1976). This is one miner's story of the arbitrary decline of a community. The Bowers Company, Altofts, West Yorkshire.

Elliott, B *Yorkshire mining veterans.* (Barnsley: Wharncliffe Books, 2005). An extraordinary collection of stories told by men whose memories span nearly a century from the early 1900's to the great strike of 1984/85 as well as the pit closures of the 1990's.

Vernon, F *Pride and poverty: memories of a Mexborough miner.* (Doncaster: Doncaster Library Service, 1984).

Other books

Linahan, L *Pit Ghosts, padfeet and poltergeists: stories and legends of the South Yorkshire area.* (Barnsley: The King's England Press, 1995). Faithful reproductions of authenticated happenings, largely in the pits of South Yorkshire. 'Read this book solely for the ghost story aspect, or use it as an unusual source for tracing pit ancestors' said the review in Family Tree Magazine.

 Linahan, L *More pit ghosts, padfeet and poltergeists: spectres, sorcery and spirits of the seam.* (Barnsley: The King's England Press, 1996).

Websites

 www.j31.co.uk/home9.html *Coal Mining Heritage.*

 www.a2a.org.uk *Access to Archives*
The archive services of South Yorkshire (Barnsley, Doncaster, Rotherham and Sheffield) have collaborated in a major project to make the existing catalogues relating to the coal industry in South Yorkshire available online.

 www.freewebs.com/yorkshiremain/ *Yorkshire Main Colliery*

The East Midlands

The Derbyshire and Nottinghamshire field was a continuation of the Yorkshire field. Centred on Ashby-de-la-Zouch, the Leicestershire and South Derbyshire field was about 50 square miles, in area.

 ***Derbyshire Record Office***, New Street, Matlock, Derbyshire
Guide to Record Office available via website.

 Sheepbridge Coal and Iron Co. accident record books 1923-1957; Denby Colliery Co. wages book 1919-1937; N.U.M. (Derbyshire Area) minutes etc.; diaries of J. F. Russ of Barlborough, coalminer, 1917-1929

 County Hall, Matlock, Derbyshire, DE4 3AG

 01629 585347

 record.office@derbyshire.gov.uk

www.derbyshire.gov.uk/leisure/record_office

Blot on the landscape?: industry and trade in the county (Guide AF25); Stone, G. *Strike action at Swanwick Colliery during the nineteenth century.*

Monthly news update in PDF format from website

Nottinghamshire Archives

Duke of Portland MSS; Colliery records (listed on website)

County House, Castle Meadow Road, Nottingham, NG2 1AG

0115 958 1634

archives@nottscc.gov.uk

www.nottscc.gov.uk

Catalogue: http://nawcat.nottinghamshire.gov.uk/

University of Nottingham, Manuscripts and Special Collections

Family and estate collections; material from former British Coal libraries (restricted); D.H. Lawrence collection; East Midlands collection

King's Meadow Campus, Lenton Lane, Nottingham, NG7 2NR

0115 9514565

mss-library@nottingham.ac.uk

www.nottingham.ac.uk/mss

Chesterfield Library

Online catalogue available

Photocopies of some Mines Inspectors' Reports; 'The Coal Collection' of books and periodicals; Derbyshire Miner 1957-1983

New Beetwell Street, Chesterfield, Derbyshire, S40 1QT

01246 209292

chesterfield.library@derbyshire.gov.uk

County Hall Local Studies Library

Reports of H.M. Inspectors of Mines; *Mining Journal* 1835-1880

County Hall, Matlock, Derbyshire, DE4 3AG

01629 585579

localstudies@derbyshire.gov.uk

www.derbyshire.gov.uk/leisure/local_studies/default.asp

Nottingham Local Studies Library

Online catalogue available

Newscuttings files on individual collieries

Central Library, Angel Row, Nottingham, NG1 6HP

0115 915 2873

local_studies.library@nottinghamcity.gov.uk

www.nottinghamcity.gov.uk

Library catalogue: http://nelib2.nottscc.gov.uk/rooms/

The Silk Mill – Derby's Museum of Industry and History

artifacts relating to the East Derbyshire coalfield

Silk Mill Lane, off Full Street, Derby, DE1 3AF

01332 255308

museums@derby.gov.uk

www.derby.gov.uk/museums

John King Workshop Museum

Various artifacts to commemorate the life and work of John King

Victoria Road, Pinxton, Nottingham, NG16 6LR

Smith, F *The life story of John King and the mine cage safety detaching hook invention.* (Pinxton: The Museum, 1970).

Snibston

Gallery illustrating the history of coalmining in Leicestershire from bell pits to the Asfordby superpit; reconstructed Tudor mineshaft; restored colliery buildings

Ashby Road, Coalville, Leicestershire, LE67 3LN

01530 510851

snibston@leics.gov.uk

www.leics.gov.uk

Useful publications:

General description of coalfield

Bell, D *Memories of the Derbyshire coalfields.* (Newbury: Countryside Books, 2006). Includes many old photographs, maps of pit areas, a chart of each colliery's life span and a list of the numbers each one employed.

Griffin, A R *Mining in the East Midlands 1550-1947.* (London: Cass, 1971). Covers mining in Nottinghamshire, Derbyshire and Leicestershire, and includes an index and bibliography.

Lawrence, D H 'Nottingham and the mining countryside' in *Phoenix: the posthumous papers of D.H. Lawrence.* (London: Heinemann, 1936). pp. 133 140.

Owen, C *Leicestershire and South Derbyshire coalfield 1200-1900.* (Ashbourne: Moorland Publishing, 1984). A very dense and detailed history of a minor coalfield which, by 1900, was already in decline.

Page, W ed *Victoria History of the County of Derby*, vol. 2. (London: Constable, 1907). pp.349-356.

Page, W ed *Victoria History of the County of Nottingham*, vol. 2. (London: Constable, 1910). pp.324-330. Includes list of collieries, pp. 328-330.

Waller, R J *The Dukeries transformed: the social and political development of a twentieth century coalfield.* (Oxford: Clarendon Press, 1983). A study of the development of new mining communities in Nottinghamshire during the 1920s and 1930s.

Social history

Griffin, A R and Griffin, C P 'A social and economic history of Eastwood and the Nottinghamshire mining country' in Sagar, K (ed.) *A D.H. Lawrence handbook.* (Manchester: Manchester University Press, 1982).

Union histories

Griffin, A R *Miners of Nottinghamshire, 1914-1944: a history of the Nottinghamshire Miners' Unions.* (London: Allen & Unwin, 1962). An unbiased picture of the struggles in the Nottinghamshire coalfield in the 1920s and 1930s which culminated in the formation of the Nottinghamshire Federated Union in 1937 and the subsequent formation of the N.U.M. - from the Foreword.

Griffin, C P *Leicestershire and South Derbyshire miners, vol 1, 1840-1914.* (Coalville: NUM Leicester Area, [1982]).

Williams, C *Derbyshire miners centenary 1880-1980.* (Chesterfield: N.U.M. (Derbyshire Area), 1980). Photographic guide to the miners of Derbyshire. A note to the Introduction is very telling – '... the greater part of the underground photographs were taken for the colliery proprietors and depict the better working conditions'.

Williams, J E *Derbyshire miners: a study in industrial and social history.* (London: Allen & Unwin, 1962). An academic history of the Derbyshire Area of the National Union of Mineworkers.

Biographies/autobiographies

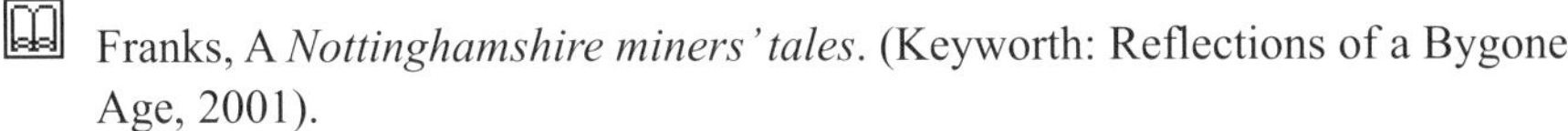 Franks, A *Nottinghamshire miners' tales.* (Keyworth: Reflections of a Bygone Age, 2001).

Judge, T *Journal of a Derbyshire pitman, 1835-1906.* (Bolsover: Highedge Historical Society, 1996. The autobiography of Joseph Wright, a Derbyshire hewer.

Woodward, M *Coalville miner's story: sixty years recollection of work in the Leicestershire coal field.* (Stroud: A. Sutton, 1993).

The West Midlands

There were a number of coalfields in the counties of Staffordshire, Warwickshire and Shropshire, more or less detached from each other and of varying sizes. North Staffordshire was about 110 square miles and South Staffordshire 150 square miles. Between Coventry, Nuneaton and Tamworth the Warwickshire field extended to about 60 square miles, and that of Shropshire and Worcestershire about 98 square miles.

Dudley Archives and Local History Service

Colliery workmen's pay books 1848-1945 (22 volumes)

Mount Pleasant Street, Coseley, Dudley, West Midlands, WV14 9JR

01384 812770

archives.centre@dudley.gov.uk

www.dudley.gov.uk/archives

Black Country History archives catalogue:
www.blackcountryhistory.org.uk/

Staffordshire Record Office

Online catalogue available

Records of North Staffordshire, South Staffordshire and Cannock Chase coalfields; NCB records

Eastgate Street, Stafford, ST16 2LZ

01785 278379

staffordshire.record.office@staffordshire.gov.uk

www.staffordshire.gov.uk/archives

Catalogue: www.archives.staffordshire.gov.uk

Guide to coal mining records

Stoke-on-Trent City Archives

North Staffordshire Miners' Federation minutes 1901-1969; North Staffordshire Mining Institute transactions 1873-1891

City Central Library, Bethesda Street, Hanley, Stoke-on-Trent, ST1 3RS

01782 238420

stoke.archives@stoke.gov.uk

www.staffordshire.gov.uk/archives

Keele University

Online catalogue available

William Jack collection – photographs, colliery minute books, scrapbooks, ephemera

Special Collections & Archives, Keele University Library, Keele, Staffordshire, ST5 5BG

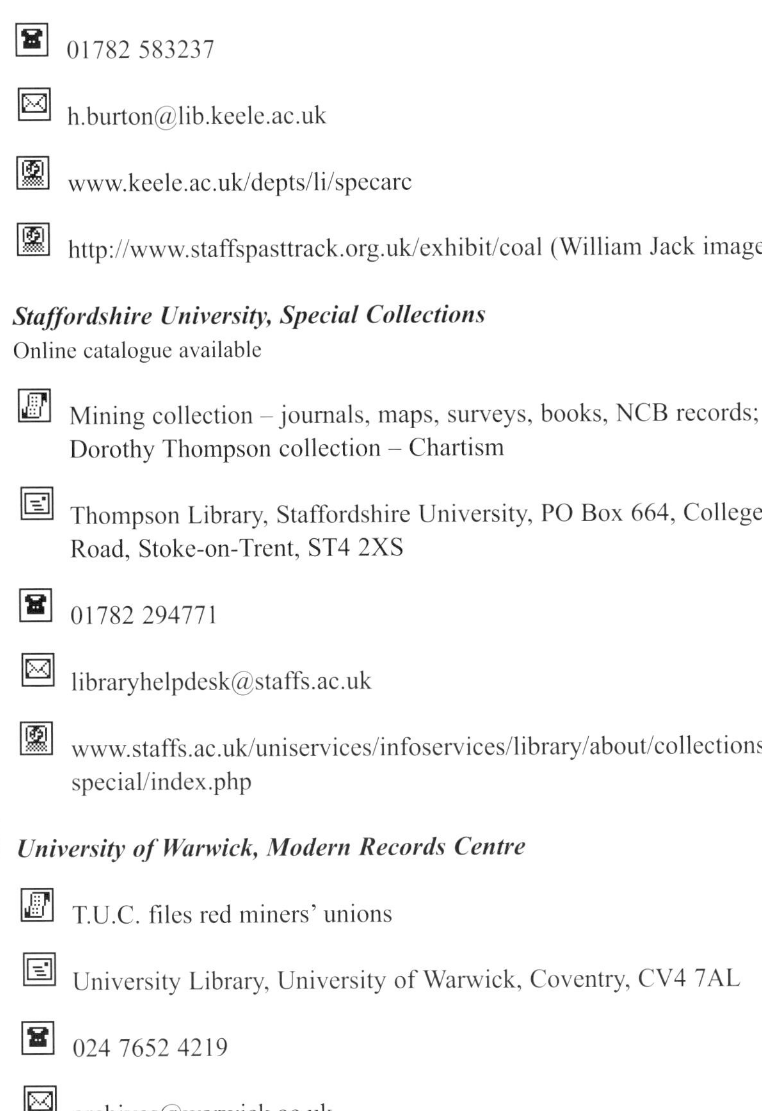

01782 583237

h.burton@lib.keele.ac.uk

www.keele.ac.uk/depts/li/specarc

http://www.staffspasttrack.org.uk/exhibit/coal (William Jack images)

Staffordshire University, Special Collections

Online catalogue available

Mining collection – journals, maps, surveys, books, NCB records; Dorothy Thompson collection – Chartism

Thompson Library, Staffordshire University, PO Box 664, College Road, Stoke-on-Trent, ST4 2XS

01782 294771

libraryhelpdesk@staffs.ac.uk

www.staffs.ac.uk/uniservices/infoservices/library/about/collections/special/index.php

University of Warwick, Modern Records Centre

T.U.C. files red miners' unions

University Library, University of Warwick, Coventry, CV4 7AL

024 7652 4219

archives@warwick.ac.uk

http://modernrecords.warwick.ac.uk

Birmingham Archives and Heritage Service

Other material in Social Sciences Department
Online catalogue available

Newscuttings on Hampstead Colliery 1875-1909; Reports of H.M. Inspectors of Mines

Central Library, Chamberlain Square, Birmingham, B3 3HQ

0121 303 4549/4217

archives.heritage@birmingham.gov.uk

www.birmingham.gov.uk

series of leaflets

Apedale Heritage Centre

Miner's cottage; artifacts; underground mine tour; above-ground coal seam

Apedale Heritage Centre, Loomer Road, Chesterton, Newcastle-under-Lyme, Staffordshire. ST5 7JS

01782 565050

info@apedale.co.uk

www.apedale.co.uk

Black Country Living Museum
Access to library by prior appointment

'Into the Thick' red-created drift mine; underground tours; Racecourse Colliery (1910)

colliery ledgers and account books

Tipton Road, Dudley, West Midlands, DY1 4SQ

0121 557 9643

info@bclm.co.uk

www.bclm.co.uk

Ironbridge Gorge Museums

Blist Hills Mine; wooden headgear; 'Great Rock Sandwich' at Jackfield Tile Museum

Ironbridge, Telford, Shropshire, TF8 7AW

01952 432 166

www.ironbridge.org.uk

Museum of Cannock Chase

On site of former Valley Colliery.
Access to archive by prior appointment

'Seven Centuries of Coal' coalmining gallery

Valley Road, Hednesford, Cannock, Staffordshire, WS12 1TD

01543 877666

museum@cannockchasedc.gov.uk

www.cannockchasedc.gov.uk/museum

Cannock Chase Mining Historical Society

Based within and allied to, but not part of, the Museum of Cannock Chase. The society has published a number of books on the collieries of the area.

www.ccmhs.co.uk/home_page.htm

Useful publications:

General description of coalfield

Greenslade, M W and Jenkins, J G eds, *Victoria History of the County of Stafford*, vol. II. (London: Oxford University Press for the Institute of Historical Research, 1967). pp. 68-107.

Page, W ed, *Victoria History of the County of Warwick*, vol. 2. (London: Constable, 1908). pp. 217-226.

Willis-Bund, J W and Page, W eds, *Victoria History of the County of Worcester*, vol. 2. (London: Constable, 1906). pp. 264-271.

Mining histories

Brown, I *East Shropshire coalfields.* (Images of England series). (Stroud: Tempus Publishing, 1999).

The Cannock Chase coalfield and its coal mines. (Cannock: Cannock Chase Mining Historical Society, 2005).

Deakin, P *Collieries in the North Staffordshire coalfield.* (Ashbourne: Landmark, 2004).

Stone, R *The collieries and coalminers of Staffordshire.* (Chichester: Phillimore, 2007).

Social histories

Leigh, F *Mining memories: a portrait of the collieries of North Staffordshire.* (SB Publications, 1992).

Biographies/autobiographies

Brown, H *Most splendid of men: life in a mining community, 1917-25.* (London: Blandford Press, 1981). An authentic autobiographical account of life in a large, tightly-knit mining community in North Staffordshire during the very harsh and difficult period following the First World War.

Websites

www.staffspasttrack.org.uk/exhibit/coal/ *Coalmining in North Staffordshire.*

www.shropshiremining.org.uk/main.shtml *Coalmining in Madeley, and much much more.*

 www.secretshropshire.org.uk/Content/Learn/Mining/Default.asp *East Shropshire Coalfield.*

www.thepotteries.org/mining/index.htm *The history of mining in Stoke-on-Trent and district.*

Gloucestershire and Somerset

The Forest of Dean coalfield in Gloucestershire lies between the River Severn to the east and the Wye Valley to the west. The Somerset and Gloucestershire field is bounded to the west by the Severn Estuary.

Bristol Record Office

Online catalogue available

'B' Bond Warehouse, Smeaton Road, Bristol, BS1 6XN

0117 922 4224

bro@bristol.gov.uk

www.bristol.gov.uk/recordoffice

Bristol University Library, Special Collections

Open by appointment only to all genuine researchers.

records of the Somerset Miners' Association 1868-1964, DM443 (34 archive boxes, 56 bound volumes)

Tyndall Avenue, Bristol, BS8 1TJ

0117 928 8014

special-collections@bristol.ac.uk

www.bristol.ac.uk/is/library/collections/specialcollections

Gloucestershire Archives

Forestry Commission offices deposit; other records relating to mining in the Forest of Dean [image] Clarence Row, Alvin Street, Gloucester, GL1 3DW

[image] 01452 425295

[image] archives@gloucestershire.gov.uk

[image] www.gloucestershire.gov.uk/archives

[image] General index of mining records

[image] ***Somerset Archive and Record Service***
Online catalogue available

[image] NCB and British Coal Corporation records; Wage books c. 1911-1950; compensation registers c. 1927-1948

[image] Somerset Record Office, Obridge Road, Taunton, Somerset, TA2 7PU

[image] appointments: 01823 337600; other enquiries: 01823 278805

[image] archives@somerset.gov.uk

[image] www.somerset.gov.uk/archives

[image] *Records of extraction industries*

[image] ***Dean Heritage Centre***
Includes the Museum (five galleries); the Gage Library; outside displays.

[image] Reconstructed Freeminer's Level, Deputy Gaveller's office, forester's cottage; paintings, models, photographs and ephemera

[image] records of the Deputy Gaveller

[image] Camp Mill, Soudley, Cinderford, Forest of Dean, Gloucestershire, GL14 2UB

[image] 01594 822170

info@deanheritagemuseum.com

www.deanheritagemuseum.com

Radstock Museum

Reconstructed coalface; miner's cottage kitchen; blacksmiths and carpenters shops; co-op shop; Victorian schoolroom; local connections with Wesley and Nelson. Temporary exhibitions three times a year.

mining records, letters and photographs – consult by appointment

Waterloo Road, Radstock, Bath, BA3 3EP

01761 437722

info@radstockmuseum.co.uk

www.radstockmuseum.co.uk

Dexter, J. *A guide to Radstock Museum.* (Radstock: Radstock Museum, 1999).

Somerset mining memories – a DVD telling the story of coalmining in Somerset by the men and women who lived through it.

Useful publications:

General description of coalfield

Down, C G and Warrington, A J *The history of the Somerset coalfield.* New ed. (Radstock: Radstock Museum, 2005). Although this coalfield was considered a minor one, this detailed history actually covers 79 collieries in the county of Somerset and is illustrated with photographs, maps and plans.

Page, W ed. *Victoria History of the County of Gloucester*, vol. 2. (London: Constable, 1907). pp. 215-238. Includes a detailed description of the history of 'the special position and privileges of the mediaeval freeminer', p. 220; and a list of collieries, p. 234.

Page, W ed. *Victoria History of the County of Somerset*, vol.2. (London: Constable, 1911). pp. 379-388. Includes list of 'chief collieries now at work in Somerset', p. 388.

Mining histories

Cornwell, J *Collieries of Somerset & Bristol.* (Ashbourne: Landmark, 2005). Recollections of the final days of the Somerset Coalfield, illustrated with many high quality photographs

Southway, M *Kingswood coal.* (Bristol: South Gloucestershire Mines Research Group, 2008).

Social histories

Anstis, R *Blood on coal: the 1926 General Strike and the miners' lockout in the Forest of Dean.* (Witney: Black Dwarf Lightmoor, 2000). A grim record of heroism and self-sacrifice based on contemporary accounts, including those of Foresters who lived through these bleak times.

Hart, C E *The free miners of the Royal Forest of Dean and Hundred of St. Briavels.* 2nd ed. (Witney: Black Dwarf Lightmoor, 2002). An updated comprehensive history of this unique group of coalminers.

Macmillan, N *Coal from Camerton: a brief history of mining in the village from 1780 to 1950.* (Avon Industrial Buildings Trust, 1990).

Biographies/autobiographies

Anstis, B and R *Diary of a working man, 1872-1873: Bill Williams in the Forest of Dean.* (Stroud: A. Sutton, 1995). A year in the life of an educated working-class man employed at the Trafalgar Colliery near Cinderford in the Forest of Dean.

Flower, F *Somerset coalmining life: a miner's memories.* (Millstream Books, 1990).

 Phelps, H *Forest voices.* (Stroud: Chalford, 1996). Recollections of local people with plenty of photos and many anecdotes from miners in the Forest of Dean.

Websites

 www.fweb.org.uk/dean/deanhist/miners.htm
History of the Forest of Dean coalfield.

 www.libswest-wisdom-sw.net/cgi-bin/crxz/index.pl
Online catalogue of many public libraries in South West England.

Kent

The coalfield is about 250 square miles on the east coast of the county of Kent. Coal was not found there until 1890, the first shaft being sunk at Shakespeare Cliff in 1896.

 ***East Kent Archives Centre***

Online catalogue available. Opened 2000, many records uncatalogued and not available.

 N.C.B. records (1911-1950); N.U.M. records; NACODS minute books (1944-1987)

 Enterprise Zone, Honeywood Road, Whitfield, Kent, CT16 3EH

 01304 829306

 eastkentarchives@kent.gov.uk

 www.kent.gov.uk

 ***Deal Library***

Online catalogue available

 newspaper cuttings, photographs, maps relating to coalfield

 Broad Street, Deal, Kent, CT14 6ER

☎ 01304 374726

✉ deallibrary@kent.gov.uk

⊞ www.kent.gov.uk

Useful publications:

General description of coalfield

📖 Ritchie, A E *The Kent Coalfield: its evolution and development.* (London: Iron & Coal Trades Review, 1922).

📖 Page, W ed. *Victoria County History of the County of Kent*, vol. 3. (London: St. Catherine's Press, 1932). pp. 380-384.

Social histories

📖 Pitt, M *The world on our backs: the Kent miners and the 1972 miners' strike.* (London: Lawrence and Wishart, 1979).

WALES

Wales had two coalfields – in the south from St Bride's Bay eastwards to Pontypool and in the north in Flintshire and Denbighshire. The South Wales Coalfield, the principal coalfield of the United Kingdom, extended for nearly 90 miles east to west, its width varying from 16 miles in the main part of the field, to a mere four miles at most in south Pembrokeshire, where the last colliery closed in 1948. Further east the proximity of the coast was very influential in the development of the coalfield. In the north, the field is much smaller, only 45 miles in length.

North Wales

🅰 *Flintshire Record Office*

🏢 Gresford Colliery Disaster Relief Fund papers 1934-1984; N.C.B. (North Wales) – various collieries; N.U.M. (North Wales) records 1891-1989

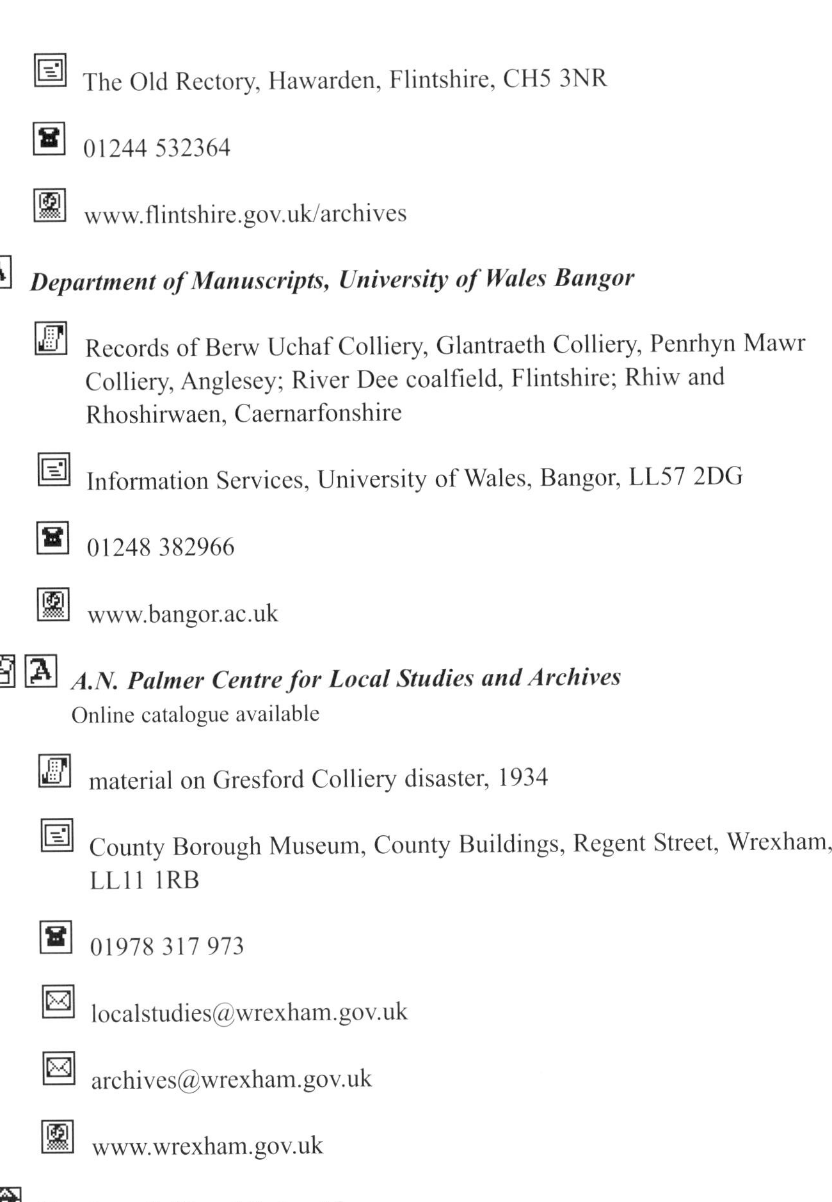

The Old Rectory, Hawarden, Flintshire, CH5 3NR

01244 532364

www.flintshire.gov.uk/archives

Department of Manuscripts, University of Wales Bangor

Records of Berw Uchaf Colliery, Glantraeth Colliery, Penrhyn Mawr Colliery, Anglesey; River Dee coalfield, Flintshire; Rhiw and Rhoshirwaen, Caernarfonshire

Information Services, University of Wales, Bangor, LL57 2DG

01248 382966

www.bangor.ac.uk

A.N. Palmer Centre for Local Studies and Archives

Online catalogue available

material on Gresford Colliery disaster, 1934

County Borough Museum, County Buildings, Regent Street, Wrexham, LL11 1RB

01978 317 973

localstudies@wrexham.gov.uk

archives@wrexham.gov.uk

www.wrexham.gov.uk

Bersham Colliery Mining Museum

The former engine house and the last headgear still standing in the North Wales coalfield.

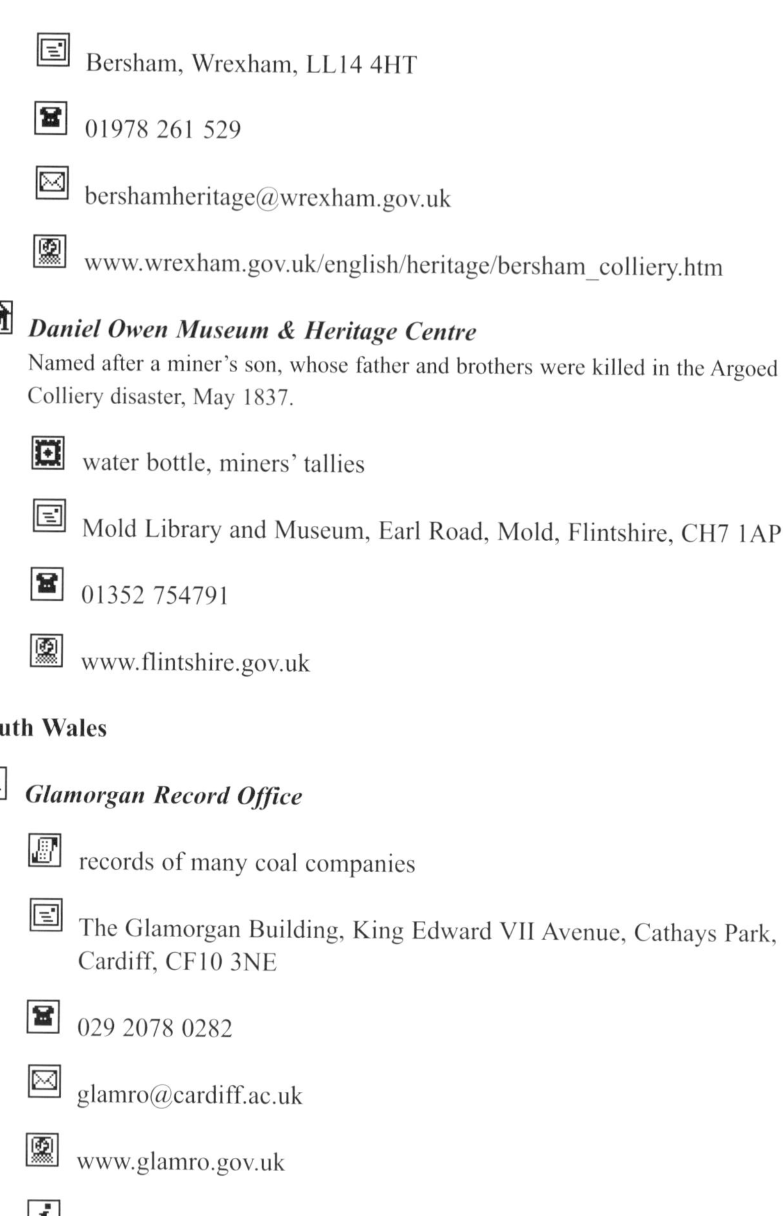

Bersham, Wrexham, LL14 4HT

01978 261 529

bershamheritage@wrexham.gov.uk

www.wrexham.gov.uk/english/heritage/bersham_colliery.htm

Daniel Owen Museum & Heritage Centre

Named after a miner's son, whose father and brothers were killed in the Argoed Colliery disaster, May 1837.

water bottle, miners' tallies

Mold Library and Museum, Earl Road, Mold, Flintshire, CH7 1AP

01352 754791

www.flintshire.gov.uk

South Wales

Glamorgan Record Office

records of many coal companies

The Glamorgan Building, King Edward VII Avenue, Cathays Park, Cardiff, CF10 3NE

029 2078 0282

glamro@cardiff.ac.uk

www.glamro.gov.uk

Colliery Records for Family Historians: Employment Records (Information Sheet no. 15)

Gwent Record Office

records of many coal companies; N.U.M. (South Wales Area) 1907-74

County Hall, Cwmbran, Gwent, NP44 2XH

01633 644886

www.llgc.org.uk/cac/

The Coal and Iron Industries (Guide to research no. 3).

Pembrokeshire Record Office

records of various collieries

The Castle, Haverfordwest, Pembrokeshire, SA61 2EF

01437 763707

www.pembrokeshire.gov.uk

Topic List: *Coal Mining in Pembrokeshire* (2004).

South Wales Miners' Library
Online catalogue available via website

South Wales Coalfield Collection – printed material, oral history, posters and banners

Hendrefoelan House, Hendrefoelan, Gower Road, Swansea, SA2 7NB

01792 518603

www.swan.ac.uk/swcc/

miners@swan.ac.uk

information leaflet

📖 D. Bevan *Guide to the South Wales Coalfield Archive.* (1974);*Supplementary guide to the South Wales Coalfield Archive*

Swansea University, Archives

Online catalogue available via website

South Wales Coalfield Collection – mss. records and photographs

Library and Information Centre, Singleton Park, Swansea, SA2 8PP

01792 295021

archives@swansea.ac.uk

www.swan.ac.uk/lis/historicalcollections/Archives/

📖 see South Wales Miners' Library

Aberdare Library

Ferndale Relief Fund minute books 1867-1886; pay books; lodge minute books; list of directors of Powell Duffryn Steam Coal Co., 1887-1944

Green Street, Aberdare, CF44 7AG

01685 885318

www.rhondda-cynon-taff.gov.uk

Newport Library

Mining Inspectors' Reports; accident reports

Reference Library, John Frost Square, Newport, NP20 1PA

01633 265539

www.newport.gov.uk

Big Pit Mining Museum

An exceptionally complete coalmining site which plays an important role in preserving the heritage of the South Wales coalfields.

Underground tour; blacksmith's shop; baths; displays of machinery etc.

Blaenafon, Gwent, NP4 9XP

01495 790311

www.nmgw.ac.uk/bigpit/index.en

www.museumwales.ac.uk/en/rhagor/article/2049 GLO annual magazine online

Cefn Coed Colliery Museum

'Coalface 1900'; winding engine; pit ponies; miners' lamps

Neath Road, Crynant, Neath, SA10 8SN

01639 750556

www.neath-porttalbot.gov.uk

Cyfartha Castle Museum & Art Gallery

Archive open by appointment

miners' lamps; mock-up of mine roadway; other mining equipment

Brecon Road, Merthyr Tydfil, CF47 8RE

01685 723112

www.museums.merthyr.gov.uk

Kidwelly Industrial Museum

steam winding engine; headframe; various colliery artifacts

Broadford, Kidwelly, Carmarthenshire, SA17 4LW

01554 891078

www.carmarthenshire.gov.uk

Museum of Welsh Life
Archive open by appointment

Audio visual, Manuscript and Photographic archives

house furnished in style of miner's family in 1895; Oakdale Workmen's Institute

St Fagans, Cardiff, CF5 6XB

0129 2057 3500

www.museumwales.ac.uk/en/stfagans

post.general@museumwales.ac.uk

Pontypool Museum

artifacts and records relating to Llanerch Colliery Disaster 1890

Park Buildings, Pontypool, Gwent, NP4 6JH

01495 752036

pontypoolmuseum@hotmail.com

pontypoolmuseum.org.uk

Rhondda Heritage Park

'Black Gold' - the story of coal through the experiences of three generations of mineworkers; 'A Shift in Time' – 'a thrilling underground experience'; reconstruction of village street; artifacts exhibitions

Lewis Merthr Colliery, Coed Cae Road, Trehafod, Rhondda,CF37 7NP

01443 682036

www.rhonddaheritagepark.com

Rhondda Museum

miners' lamps and equipment; wages ledgers

Glyncornel House, Nanty Gwyddon Road, Rhondda, CF40 2JR

www.rhondda-cynon-taff.gov.uk

South Wales Miners' Museum

various displays and mining artifacts

Afan Forest Park Visitor Centre, Cynonville, Port Talbot, SA13 3HG

01639 850564

www.npt.gov.uk

afanforestpark@npt.gov.uk

Winding House
Research room on site

winding house and steam winding engine

Cross Street, New Tredegar, NP24 6EG

☎ 01443 822666

✉ windinghouse@caerphilly.gov.uk

▦ www.caerphilly.gov.uk/windinghouse/english/home.html

Useful publications:

General description of coalfield

📖 Williams, G ed *Glamorgan County History*, vol. v. 'Industrial Glamorgan from 1700 to 1970', (Cardiff: Glamorgan County History Trust Ltd., 1980). pp. 155-209.

Mining histories

📖 Carpenter, D J *Rhondda Collieries* (Images of Wales series). (Stroud: Tempus Publishing, 1999).

📖 Cornwell, J *Collieries of South Wales,* 2 volumes. (Ashbourne: Landmark, 2001, 2002).

📖 Cornwell, J *Rhondda Cynon Taff Collieries.* (Ashbourne: Landmark, 2008).

📖 Jacob, C *Coal mining in Merthyr Tydfil and District.* (Merthyr Tydfil: Merthyr Tydfil Public Libraries, 1997). A brief history, illustrated with evocative photographs.

📖 Kelly, I *North Wales coalfield: a collection of pictures.* vol. 1. (Bridge Books, 1990).

📖 Rees, R *The black mystery: coalmining in South Wales.* (Talybont:Y Lolfa, 2008).

📖 Symons, M V *Coal mining in the Llanelli area, vol. 1, 16th century to 1829.* (Llanelli: Llanelli Borough Council, 1979). Detailed history of a very specific area and contains a useful schedule of pits.

Social histories

Bellamy, D *Images of South Wales mines*. new ed. (Stroud: Sutton, 1997).

Evans, R M *Children in the mines 1840-2*. (Cardiff: National Museum of Wales, 1972).

Evans, R M *Children working underground*. (Cardiff: National Museum of Wales, 1979).

Eyers, M *The masters of the coalfield: people and place names in Glamorgan and Gwent*. (M. Eyers, 1991). A fascinating study of placenames in the mining areas of Glamorgan and Gwent, and the people after whom they were named.

Fieldhouse, B and Dunn, J *'And they worked us to death'*. 3 vols. (Gwent F.H.S., 1992, 1993, 2002). Provides details of mining accidents that occurred in the valleys of South Wales from the nineteenth century onwards, with lists of names of casualties.

Rhondda Borough Council *'Black diamond': to mark the closure of the last pit in the Rhondda*. (Rhondda Council, 1991).

Williams, R and Jones, D *The cruel inheritance: life and death in the coalfields of Glamorgan*. (Griffithstown, 1990).

Williamson, S *Gresford: the anatomy of a disaster*. (Liverpool: Liverpool University Press, 1998).

Union histories

Arnot, R P *South Wales miners (Glowyr de Cymru): a history of the South Wales Miners' Federation 1898-1914*. (London: Allen & Unwin, 1967). A history of the 'formation and often turbulent years of the South Wales Miners' Federation' - foreword.

Arnot, R P *South Wales miners, 1914-1926*. (London: Allen & Unwin, 1975).

 Francis, H and Smith, D *The Fed: a history of the South Wales Miners in the twentieth century.* (London: Lawrence & Wishart, 1980). Although published as the official history of the union, this is more than that – more a social history of the area, illustrated with maps and photographs, which bring the story to life.

Biographies/Autobiographies

 Coombes, B L *These poor hands: the autobiography of a miner in South Wales.* (Cardiff: University of Wales Press, 2002).

 Ellis, T *After the dust has settled.* (Wrexham: Bridge Books, 2004). The autobiography of a coalminer, manager and politician whose parliamentary career coincided with the political upheavals of the 1970s and led him to campaign long and hard for electoral reform and Britain's greater involvement in the European Union.

Websites

 www.bbc.co.uk/wales/coalhouse/index1.shtml *BBC Wales' 'Coalhouse'.*

 www.bbc.co.uk/wales/history/sites/themes/society/family_01_gettingstarted .shtml *BBC Wales guide to family history*

 www.therhondda.co.uk/intro.html *The Rhondda*

www.welshcoalmines.co.uk *Welsh Coal Mines*
Informative site that includes a 'forum' for queries about Welsh mining ancestors.

SCOTLAND

There were several important coalfields in Scotland extending from the Firth of Forth to the Firth of Clyde. This zone is about 95 miles long, and its width varies up to 30 miles. The principal fields were the Midlothian, Fifeshire, Linlithgow and Lanarkshire, Clackmannan and Ayrshire fields.

The Central Scottish coalfield extended from Paisley and Barrhead in an easterly direction for some thirty miles to Bathgate. From south to north it extended from East Kilbride for about 21 miles into the Falkirk area. There was a north easterly extension into Fife. The Ayrshire field lay on the east shore of the Firth of Clyde and was about 330 square miles in area. The smaller Midlothian field is geologically the southern end of the Fife and Forth basin, and was probably the oldest worked in Scotland.

The earliest records of the working of coal in Scotland relate to the Midlothian coalfield in the twelfth century, when the monks of Newbattle worked coal on the banks of the Esk; and to Linlithgow, where the monks of Holyrood had, about the same time, received a grant of land on which coal was dug.

Ayrshire Archives
The joint archive service of East, North and South Ayrshire. Includes the Ayrshire Sound Archive.

Sound Archive includes personal recollections of various Ayrshire coalminers

Ayrshire Archives Centre, Craigie Estate, Ayr, KA8 0SS

01292 287584

archives@south-ayrshire.gov.uk

www.ayrshirearchives.org.uk

National Archives of Scotland
Online catalogue available

Records of early coal companies; NCB records

West Register House, Charlotte Square, Edinburgh, EH2 4DF

0131 535 1314

wsr@nas.gov.uk

www.nas.gov.uk

www.nas.gov.uk/guides/coalmining.asp Coal mining records

The coalminers. (Edinbugh: Scottish Record Office, 1983).

A *National Library of Scotland*

Records of the National Union of Mineworkers, Scotland, and its branches; the original drafts of R. Page Arnot's *History of the Scottish Miners* (1955); correspondence and papers of Willie Hamilton, former M.P. for Fife.

George IV Bridge, Edinburgh, EH1 1EW

0131 226 4531

www.nls.uk

Dunfermline Central Library

Reports of H.M. Inspectors of Mines 1854-1919, 1930, 1950
'Twentieth century Dunfermline' project includes mining-related video and audio material

1 Abbot Street, Dunfermline, Fife, KY12 7NL

01383 602365

www.fifedirect.org.uk

A *East Dunbartonshire Libraries*

photographic collection; records of local coal companies

William Patrick Library, 2-4 West High Street, Kirkintilloch, G66 1AD

0141 775 4574

libraries@eastdunbarton.gov.uk

www.eastdunbarton.gov.uk

Edinburgh Central Library

Reference Library, Scottish Department, and Edinburgh Room all have items of interest.

N.U.M. (Scottish Area) Newcraighall Branch minutes 1940-1968

George IV Bridge, Edinburgh, EH1 1EG

0131 242 8000

edinburgh.room@edinburgh.gov.uk

www.edinburgh.gov.uk

Kirkcaldy Museum and Art Gallery

Archive open by appointment.

small permanent display on coalmining in Fife; table and chairs carved from 'Parrot' coal

Abbotshall Road, Kirkaldy, Fife, KY1 1YG

01592 583213

kirkcaldy.museum@fife.gov.uk

www.fifedirect.org.uk

Scottish Mining Museum

Library open by appointment.

Archive of the Lothian Coal Company; over 14,000 photographs

'The Story of Coal'; 'A Race Apart'; 'Big Stuff' tour

Lady Victoria Colliery, Newton Grange, Midlothian, EH22 4QN

☎ 0131 663 7519

✉ enquiries@scottishminingmuseum.com

▦ www.scottishminingmuseum.com

🏛 *Summerlee – Museum of Scottish Industrial Life*

Reopened in 2008 after a major refurbishment

▣ Cardowan Colliery and Farme Colliery winding engines; drift mine; miners' cottages with interiors 1840-1960

▤ Heritage Way, Coatbridge, ML5 1QD

☎ 01236 423256

✉ museums@northlan.gov.uk

▦ www.northlan.gov.uk

Useful publications:

Family history

📖 Durie, B *Scottish genealogy*. (Stroud :The History Press, 2009). A thorough and up-to-date guide to Scottish genealogy

📖 Reeks, L S *Scottish coalmining ancestors*. (Baltimore: Gateway Press, 1986). Not quite as useful as the title suggests, nevertheless this book is based on extensive research based on the records of the parish of Newton, in which lies the Scottish Mining Museum.

Mining histories

📖 Arnot, R P *History of the Scottish miners from the earliest times*. (London: Allen & Unwin, 1955). The story of mining in Scotland from Roman times; the 'slavery' of the Scottish miner in the seventeenth and eighteenth centuries; and including a vivid account of the Knockshinnon disaster of 1950 and its aftermath.

Campbell, A *The Scottish miners, 1874–1939*, 2 vols. (Aldershot: Ashgate, 2000). These books represent not just the study of one group of workers and their families but also make an important and novel contribution to the social history of modern Scotland.

Duckham, B F *History of the Scottish coal industry, vol. 1., 1700-1815.* (Newton Abbot: David & Charles, 1970). Subtitled 'a social and industrial history', this book describes the technical, economic and social development of coalmining in Scotland. It provides an informative background to the industry in Scotland.

Oglethorpe, M K *Scottish collieries: an inventory of Scotland's coal industry in the nationalised era.* (Edinburgh: RCAHMS, 2006).

Social history

Duncan, R *The mineworkers* (Edinburgh: Birlinn, 2005). Concentrates on the work, working conditions and struggles of generations of men, women and children who laboured underground and at the pit-head in the principal mining areas of west, central and eastern Scotland.

Halliday, R S *The disappearing Scottish colliery.* (Edinburgh: Scottish Academic Press, 1990). A personal farewell to almost 200 Scottish collieries and their communities.

Hutton, G *Fife: the mining kingdom.* (Catrine: Stenlake, 1999).

Hutton, G *Lanarkshire's mining legacy.* (Catrine: Stenlake, 1997).

Hutton, G *Mining: Ayrshire's lost industry.* (Catrine: Stenlake, 1996).

Hutton, G *Mining: the Lothians.* (Catrine: Stenlake, 1998).

M'Neil, P *Blawearie, or Mining life in the Lothians forty years ago.* (Leicester: Remploy, 1983). First published 1887. McNeill was a bookseller, born in 1839. His grandfather was one of the last miners to work as a serf and his mother worked in the pits until 1848.

Scottish Record Office *The coalminers*. (Edinburgh: Scottish Record Office, 1983). A large format booklet with many evocative photographs of archive material including the Loanhead, Midlothian, Colliers' Sickness Fund, 1720, on which only 5 out of 15 colliers are able to sign their own names.

Union histories

Campbell, A B *Lanarkshire miners: a social history of their trade unions*. (Edinburgh: J. Donald, 2003).

Biographies/autobiographies

Owens, J ed. *Miners 1984-1994*. (Edinburgh: Polygon, 1994). A thoughtful introduction is followed by the personal memoirs of miners and their families in the Scottish Coalfield – many in 'Scots' dialect.

Terris, I *Twenty years down the mines*. (Catrine: Stenlake, 2001). A first-hand miner's account of everyday work and woes.

Websites

www.bbc.co.uk/scotland/history/scotlandonfilm/media_clips/index_topic.shtml?topic=work&subtopic=mines *Scotland on Film*. Evocative film and sound clips of coalmining life in Scotland in the twentieth century.

www.coalcollections.org *Database of coal mining collections held in museums, libraries and archives throughout Scotland*

www.fifeminingheritage.org.uk *Fife Mining Heritage Society*.

www.users.zetnet.co.uk/mmartin/fifepits/ *Coalmining in Fife*.

CHAPTER 13
Printed and digital sources

PRINTED

The literature of coalmining is extensive and this chapter is not designed to be a complete guide to the thousands of publications which, over the years, have covered its historical, technical, sociological, or political aspects.

Bibliography

Benson, J *et al* (compilers) *Bibliography of the British coal industry: secondary literature, parliamentary papers, mineral maps and plans and a guide to sources.* (Oxford: Oxford University Press, 1981). A companion volume to the standard History of the Coal Industry (see below), and provides a wealth of bibliographical information. Rather outdated now, but unlikely to be replaced.

Neville, R G and Benson, J 'Labour in the coalfields (II)', *Bulletin of the Society for the Study of Labour History*, no. 31, Autumn 1975. pp. 45-59. An update of the original bibliography by J.E. Williams.

Richardson, R C and Chaloner, W H *British economic and social history: a bibliographical guide*. 3rd. ed. (Manchester: Manchester University Press, 1996). Arranged by period and country. Lists over thirty titles in the section 'England 1700-1980: Coal', plus others on trade unions and coalmining.

Williams, J E 'Labour in the coalfields: a critical bibliography', *Bulletin of the Society for the Study of Labour History*, no. 4, 1962. pp.24-32.

General literature: Classics

Galloway, R L *Annals of coal mining and the coal trade*. 2 vols. (Newton Abbott: David & Charles, 1971), first published 1898 and 1904. Volume one takes us from 'previous to 1066 AD' to 1835 and volume two continues the story up to 1850. The reprint of vol. 1 contains a useful introduction by B.F. Duckham in which he describes this as 'one of the great classics'.

Galloway, R L *A history of coal mining in Great Britain*. (Newton Abbott: David & Charles, 1969), first published 1882. A highly readable account of the general development of the industry. Duckham's introduction takes the story up to World War I.

Jevons, H S *The British coal trade*. (Newton Abbott: David & Charles, 1969), first published 1915. This work was virtually complete before the outbreak of the First World War, when coalmining dominated the industrial scene, and is essential reading for anyone interested in the British coal industry at its productive zenith.

Nef, J U *The rise of the British coal industry*. 2 vols. (London: F. Cass, 1966), first published 1932. A scholarly and detailed study of the industry from 1550 to 1700.

General literature: Modern histories

Anderson, D *Coal: a pictorial history of the British coal industry*. (Newton Abbot: David & Charles, 1982). A readable introduction to the history of the industry, fully illustrated throughout. Includes a glossary and very useful chronology.

Arnot, R P *The miners: a history of the Miners' Federation of Great Britain*. (London: Allen & Unwin, 1949).

Arnot, R P *The miners: years of struggle: a history of the Miners' Federation of Great Britain (from 1910 onwards)*. (London: Allen & Unwin, 1953).

Arnot, R P *The miners in crisis and war: a history of the Miners' Federation of Great Britain (from 1930 onwards)*. (London: Allen & Unwin, 1961).

Arnot, R P *The miners: one union, one industry: a history of the National Union of Mineworkers, 1939-46*. (London: Allen & Unwin, 1979).

Benson, J *British coalminers in the nineteenth century: a social history*. (Dublin: Gill and Macmillan, 1980). A wide-ranging survey of the miners' way of life in the nineteenth century.

Burton, A *The miners*. (London: Andre Deutsch, 1976). The story of the men, women and children who laboured under the earth digging Britain's coal; illustrated with many evocative photographs.

Griffin, A R *The British coalmining industry: retrospect and prospect*. (Buxton: Moorland Publishing, 1977). A highly readable and well-illustrated book which is one of the best single-volume histories.

Griffin, A R *The collier*. (Princes Risborough: Shire, 1982). A well-illustrated history from the point of view of the workers in the industry.

Hayes, G *Coal mining*. (Princes Risborough: Shire, 2000). Nicely illustrated and up-to-date booklet with a list of places to visit.

History of the British Coal Industry, 5 volumes: The standard academic history of the industry – authoritative but dry. Each volume has an extensive bibliography and list of archive sources.

Hatcher, J *vol. i, before 1700*. (Oxford: Clarendon Press, 1993);

Flinn, M W and Stoker, D *vol. ii, 1700-1830: the Industrial Revolution*. (Oxford: Clarendon Press, 1984);

Church, R *vol. iii, 1830-1913: Victorian pre-eminence*. (Oxford: Clarendon Press, 1986);

Supple, B *vol. iv, 1913-1946: the political economy of decline*. (Oxford: Clarendon Press, 1987);

Ashworth, W *vol. v, 1946-1982: the nationalised industry*. (Oxford: Clarendon Press, 1986).

Mining Association of Great Britain *Historical review of coal mining.* (London: Fleetway Press, 1925). History of the technical and legislative aspects of the industry, well illustrated.

Thornes, R *Coal* (Images of industry). (Swindon: R.C.H.M.E., 1994). 120 photographs of structures associated with the coal industry taken in the early 1990s.

Vernon, R W ed *Mining heritage guide.* 3rd. ed. (Keighley: National Association of Mining History Organisations, 2000). Guide to sites, museums and societies, and includes 'A brief guide to primary sources for mining history', by Peter Claughton.

For current lists of books for the family historian (which are also too numerous to list here) see the following website: www.ffhs.co.uk

DIGITAL

Websites

The number of non-printed sources increases at a phenomenal rate. Technology has made it possible to access data from repositories all over the world, in a digitised format, and in various media – microfiche, CD ROM or the Internet. Reference books, maps, illustrations, manuscript records and indexes – data of all kinds – are all available over the world wide web.

Genealogist Nick Barratt suggests that family history is the third most popular activity on the web, with some websites reporting huge increases in usage in recent years. Internet shopping is also becoming ever more popular as security systems improve, and the Genfair site gives access to much genealogical material.

www.guardian.co.uk/lifeandstyle/2008/dec/27/tracing-family-history

Many organisations also have their own website, which should be the source of the most up-to-date information. Individuals, too, have set up their own sites, which may include information on their family history, and perhaps mining history also. This book could not have been written without the web, and web addresses have been quoted throughout, wherever possible. Most websites have useful 'Links' to related sites worldwide. However the speed of change in this area is such that, during the revisions of this book before publication, the author often found that addresses were no longer valid or that links failed.

Often, useful material is to be found in unlikely places. A website, for instance on the history of Conisbrough Castle includes an evocative description of 'The worst village in England' – the Yorkshire mining village of Denaby Main – in 1899. The website of the National Mining Memorabilia Association includes not only pictures of the New Hartley Rescue Medal, but also a description of the 1862 Hartley disaster; a list of the rescuers; and links to related sites.

Some material is available on a 'charged-for' or 'pay per view' basis, but sometimes that which formerly was charged for is now available for free. An important example of this is the Coal Mining History Resource Centre website where the scanned sections of the Children's Employment Commission *Report* (see page ?) are now freely available. Similarly the website of the journal *Midland History*, previously on subscription, now includes almost all its published articles (some on coalmining) as scanned PDF documents. Another surprise was the website 'England's past for everyone'. This includes photographs of Bolsover (Derbyshire) and Wearmouth (Durham) collieries, together with articles on various aspects of the history of both collieries. A search for 'coal' returned 46 matches.

www.conisbroughcastle.org.uk/Education/socialhistory.htm
www.cmhrc.co.uk/site/literature/royalcommissionreports/
www.midlandhistory.bham.ac.uk/
www.englandspastforeveryone.org.uk
www.mining-memorabilia.co.uk

Any attempt at a comprehensive listing would be out-of-date even before publication, but guides such as *Genealogist's internet* are invaluable. For the most up-to-date information on new websites check out the various monthly genealogical magazines.

Christian, P *Genealogist's internet*. (London: TNA, 2005).

Raymond, S A *Family history on the web: an Internet directory for England and Wales 2008-2009*, 5th ed. (Bury: Family History Partnership, 2008).

There are many general coalmining sites which are likely to be of interest to readers of this book. The Mining History Network's Web Resources Page is the premier jumping off point for links to the major mining history related sites in the UK and abroad.

www.exeter.ac.uk/~RBurt/MinHistNet/ *Mining History Network*
www.minersadvice.co.uk/index.htm – general mining site
www.bbc.co.uk/nationonfilm/topics/coal-mining/ *Nation on film: coalmining*
www.namho.org/ *National Association of Mining History Organisations*
http://mininghistory.thehumanjourney.net/ *Access to mineral heritage*

Mailing lists

Mailing lists are a useful source of information and an efficient means of keeping in touch with like-minded people around the globe, and readers will find the Coalminers list of interest, although there is an American bias. There are many local and regional genealogical mailing lists – all coalmining counties appear to have one – and there is a list on GENUKI. A useful list on mining topics is the Mining History list, and this is also archived.

http://archiver.rootsweb.com/th/index/COALMINERS
www.genuki.org.uk/indexes/MailingLists.html
http://www.jiscmail.ac.uk/cgi-bin/wa.exe?A0=mining-history&T=0

Web rings

These are groups of, and direct links between, websites, and are listed at: http://w.webring.com.

Those on mining include the *Mining and Minerals Industry; History of Mining and Mineral Exploration; Mining;* and *Industrial Archaeology and History.*

CD ROMs

Many commercial organisations, and family history societies, produce interesting and valuable material on CD ROM. Available items include: Census Enumerators' Books; historical local and county directories; topographical guides; maps and plans; and county histories for coalmining areas. Increasingly, DVDs are also being used as a medium for this material.

The websites of such companies as Genfair, Parish Chest and S & N Genealogy Supplies provide an overview of available products.

www.genealogysupplies.com
www.genfair.co.uk
www.parishchest.com

APPENDIX I

Glossary

The terminology of mining has changed during the history of the industry. Although miners 'were still down the same dark holes in the ground doing in essence the same filthy work'[1], and still used many of the same terms, their meaning altered as mining techniques changed. There were differences between, and even within, the coalfields – an object could have another name in another part of the country; or the same word could have different meanings in different coalfields. There was also the problem of dialect and pronunciation. In Durham the miner's language was known as 'pitmatic', a word incorporated in the title of a reprint of one of the sources of this glossary.

adit	A tunnel driven into a hillside in connection with mineral working, for transport, ventilation or drainage (or all three).
after damp	A mixture of the deadly poisonous gases carbon dioxide and carbon monoxide which occurs after an explosion of methane.
agent	An official senior to a colliery manager, usually responsible directly to the owner.
back shift	The afternoon shift.
backbye	The areas of the mine away from the face.
baff	The alternate, or 'off' day or week when the fortnightly wages were not paid to miners.
bank	The colliery surface.
bank work	An early and primitive system of long-way working practised in Yorkshire, Derbyshire, Nottinghamshire and Leicestershire.

| *banksman* | The man in charge on the surface, responsible for raising and lowering the cage. |

banksman | The man in charge on the surface, responsible for raising and lowering the cage.

bannickers | In Yorkshire, shorts. *See* hoggers.

barrier | A solid pillar of coal left to mark the boundaries between various owners' royalties; or to prevent the migration of water or gas from old workings.

base | The estimated limit of annual output, set by coalowners' cartel in order to regulate markets

basset | The outcrop of a seam at the surface.

baulk | Wooden beam.

beans | Small coals.

bearmouth | In Cumberland, a drift mine.

bell pit | A shallow shaft, where coal is worked around the pit-bottom until the sides are in danger of collapsing when it is abandoned. Seen in section, it has the general shape of a bell.

Bevin boy | Conscripted miner in World War II. (See Chapter 7).

big butty system | Where a pit was let to one (or several) superior workmen who provided the circulating capital, paid the men and effectively managed the works, being paid so much per ton.

big hewer | The almost legendary miner who could work as hard as ten ordinary men.

binding | Agreement of pitmen with employers for one year's hire. See bond.

black-damp | see stythe.

blower | A violent discharge of firedamp.

boll | A coal measure.

bond | The document by which miners were bound to their employers for a year at a time in Northumberland and Durham (and Scotland, too, after the abolition of serfdom).

bord-and-pillar | North Eastern variant of stall-and-pillar. The bords were rectangular excavations separated by solid pillars of coal left to support the roof and making the working area resemble a chess board

brakeman/ brakesman | The man employed to work the winding-engine.

brattice | Canvas used to redirect air around the pit.

bull week	A term of Yorkshire origin. The week before a holiday when men maximised their earnings by working as many shifts as possible.
butterfly	A safety device used in the shaft.
butty	In Yorkshire, the Midlands and North Wales, a sub-contractor.
cage	A steel platform with sides and a roof in which men, materials and coal are drawn up and down the shaft.
caller	The man employed to call from house to house to wake the miner.
cannel coal	A shiny coal which gives a good light when burning.
cap	The blue top on a candle or lamp when it burns in a mixture of fire-damp and air.
cavil	System of drawing lots to determine working places.
chaldron	A measure of capacity used in the North East. A Newcastle chaldron was approximately 53 cwt. and a London chaldron about 27 or 28 cwt.
charter master	Staffordshire name for a big butty.
check-	Name for both the owner's and the miner's representative, each weighman appointed to check the other's dishonesty, in weighing coal-laden tubs, as they come from the pit.
chokedamp	A synonym for blackdamp.
chummins	Empty tubs.
cleat	The lines of cleavage of coal, analogous to the grain of timber.
coal	Solid, usually brown or black, carbon-rich material that most often occurs in stratified, sedimentary deposits. It is one of the most important of the primary fossil fuels.
coalrunner	A term used in Yorkshire for the lads who go down with the tubs.
coke	Cinders produced by burning coal.
collier	The skilled man who worked at the coal face and actually got the coal.
consideration	Extra payments made to pieceworkers supposedly to compensate for abnormal working conditions.
corf or corve	A wicker-work basket for carrying coals from the face, gradually replaced in the nineteenth century by tubs.
corporal	Man in charge of the underground haulage hands.

coup	An exchange of cavils (q.v.).
coupler	A man or boy who worked on the haulage system coupling tubs together.
coursing	A more efficient conducting of air through the mine by means of partitions and stoppings
cow	A long iron rod fastened to the last 'tub' of a 'set', so that in case the rope breaks, the rod sticks in the ground and holds the tub fast.
cracket	Stool, usually three-legged, used in some districts by hewers either to sit on or to support their shoulders, neck or elbow.
crake	The crier's rattle, used when a meeting of miners is cried through the street.
creep	The heaving up of the floor of the roadways underground, otherwise called 'floor lift'.
crib or curb	Segmental wooden rings in shafts on which tubbing or brickwork rests.
crook	An iron hook on a chain used to fasten the 'guss' (hempen harness worn by boys) to the 'put' (box of coal) in Somerset.
crosscut	An excavation driven in any direction.
crown-tree	Plank used to support the roof in coal workings.
crush	Convergence of the strata.
cupola	Ventilation furnace chimney, sometimes called a 'cube' or 'tube'.
darg	In Scotland, a day's hewing task, elsewhere called a 'stint'.
Davy	The safety lamp invented by Sir Humphrey Davy in 1815 where the flame was insulated by a wire gauze.
day-hole	A small exit vent in a drift mine, or the drift mine itself.
delf/delph	A Lancashire term to describe a seam of coal.
deputy or deputy overman	Official employed in a supervisory capacity with responsibility for setting props and general safety matters.
detaching	A device by which an overwound cage is detached from the rope and
hook	is held firmly in the headgear.
dilly line	Railway line used on haulage systems on the surface and underground.
districts	Name given to sub-divisions of the seam.

doggy	Man in charge of the haulage lads; also the wagonwayman.
downcast	The shaft down which the air flows.
draw	To remove supports which are no longer required; or, the passage of the cage through the shaft.
drawer	The person who is employed by the collier to take full tubs from the workplace to the haulage and bring back empty tubs.
drift	A heading or tunnel.
drift mine	A mine is driven from the surface (as opposed to a shaft mine).
driver	Boy who led horse-drawn full tubs along the main underground roads out-bye.
drop pit	A side tunnel into the shaft. Usually part of the ventilation system.
duff	Fine coal, or coal dust.
dyke	Considerable natural interruption of the bed of coal, whereby it is either thrown down or up.
elephant feet	The base plates attached to hydraulic props.
endless rope haulage	A long rope or chain that is driven by an engine round two pulleys, one at each end.
engineman	The man in the engine house who is in charge of the winding engine.
fault	A fracture of a coal seam caused by earth movement; is called an up throw (where the seam continues at a higher level) or a down-throw (where it continues at a lower level).
fiery heap	The deposit of rubbish and waste or unsaleable coal which usually takes fire spontaneously.
filler	A man employed on filling coal at the face.
fire bucket or lamp	Bucket like a night-watchman's brazier placed or suspended in an upcast shaft to aid ventilation.
fire clay	A refractory clay or shale.
firedamp	Inflammable gas whose chief constituent is methane. It is emitted from the seam (and also the floor and roof) during working.
fireman	An official who is in charge of a district in the mine.
fitter	Coal broker or agent.
foals	Young putters. *See* putter.

fore-shift	Any shift starting after midnight.
fother	A measure of coals.
foul air	Air in an inflammable condition arising from a concentration of fire damp gas
foul tubs	Containing quantities of coal or stone considered unfit for market
fullins/fulluns	Full tubs
galloway	Pony.
ganister	A very hard fire-clay.
gate	Underground roadway or tunnel.
gaudy	A holiday.
Geordy	The safety lamp invented by George Stephenson in 1815.
getter	A man producing coal at the face. In many cases he was also the filler (q.v.).
girder leader	The modern equivalent of the hewer (q.v.)
glenny	A safety lamp; probably a corruption of Clanny, after Dr W. R. Clanny who invented several lamps, the first in 1813.
goaf (or gob)	The waste area from which coal has been removed; it is partly filled with small coal and debris.
gob fire	Spontaneous combustion of small coal in the gob.
grove	A drift mine.
gullet	A fissure in the strata, generally containing either water or inflammable air.
hack	A heavy pick, wighing about 7 lbs., with head about 18 in. in length.
half marrows	Young putters. *See* putter.
hand money	Extra money paid at binding to encourage and consolidate agreements.
hanger-on	The man who used to hang corves on the rope in the pit-bottom. Also *(or hooker-on)* called an 'onsetter'.
haulage	The system by which the coal is transported from the coal face to the bottom of the shaft. *See* Endless Rope Haulage.
headgear	See headstocks.
heading	A tunnel driven into the coal.
headsman	A young putter.

heap/ heapstead	All the surface installations of the mine.
helper-up	The boy who assists the barrow-men, and putters.
hew	To hack away at the coal down a mine. Hence, 'hewer'.
hitch	A considerable interruption of the bed of coal.
hoggers	Stockings with the feet cut off; or, pit shorts.
holing	Undercutting the face of coal. Hhence 'holer', the man carrying out this operation.
horny or horney tram	Flat tub without sides with an upright at each corner.
hunkers	Haunches. Miners' squat in which he sits on the toes, with the thighs resting on the calves.
inbye	Towards the coal-face (as opposed to outbye or backbye – away from the face).
intake	Airway along which fresh air is taken into the workings, as opposed to the 'return' airway carrying foul air away from the workings to the upcast shaft.
jet	A coarse cannel coal.
jig	A self-acting inclined plane.
jigger	The person, usually a lad, who operated the jig; in more modern times, a small pneumatic drill.
jowl	To sound the roof.
jud	The amount which a strong man could bring down and fill in a shift.
jumper	A tapered round iron bar with a sharp point formerly used for drilling shot holes manually, especially in the Midlands.
keeker	In Northumberland and Durham, a surface foreman. The owner's 'weighman'.
keel	Boat holding about 21 tons of coal formerly used on the Tyne and Wear for carrying coal from the staithes to the collier brigs.
keeps/keps or catches	Supports which hold the cage in place at the mouth of the shaft.
kibble	A square wooden tub.
kirving	Wedged-cut into the lower half of the coal seam made by the hewer.
kist	In Northumberland and Durham a deputy's tool chest; later, his underground 'office'.

kit	A small tub for washing in.
knocking	A system of signalling used to control the cage and the haulage system.
laid out	A corf or tub of coals containing an excessive quantity of dirt, for which the hewer was fined.
lamp cabin	The surface building in which the lamps are housed.
landsale	Coal sold for transport by road (as distinct from sea-sale, canal-sale or rail-sale.)
lay in	To 'lay in' a pit, or lay it idle; to leave off working it, as when it becomes exhausted of coals.
layering	Firedamp, being lighter than air, sometimes forms a layer near the roof of a roadway, and it may move in the opposite direction to the ventilation if the current of air is weak.
level	A horizontal road in the mine; or, a water course, sometimes used also for conveying coal.
limbers/ limmers	The shafts which attached the pony to the tub or tram.
linesman	A man who puts on survey lines, a surveyor's assistant.
linings	Pitmen's drawers, fastened at the knee by strings.
longwall	A system of working coal originating in Shropshire where a number of men work along a coal-face. Roadways are maintained through the 'gob', and there are no pillars with this system.
long-way	A generic term covering longwall and various forms of bank work.
low	A light (eg candle).
main road	The main roadway driven from the pit-bottom through the workings.
man-holes	Refuge holes made along a roadway where men can shelter if, for example, a set of tubs becomes de-railed.
manriding set	Underground passenger train.
marra/marrow	In Northumberland and Durham especially, a work mate; a member of a team paid on a common pay note.
master	The old name for the owner of the pit.
master note	The common pay note paid to the members of a team.
master shifter	Head man on the night shift.

meeting	The time when the cages meet each other in the shaft.
metal man	A man employed to service the tunnels and rails in the haulage system.
midgy	Also called a 'Mistress'. These names were given to a kind of lamp used by putter lads. Originally, they were little wooden boxes, with a hole at the bottom, through which the candle was thrust, and another hole at the top to let the heat out.
mothergate	The central tunnel out of which all coal comes.
mouthing	Entrance to a seam or tunnel.
mullergate	*see* mothergate.
neuk	The extreme ends of the face.
nick	In bord-and-pillar, the hewer first undercut the coal, then cut a vertical 'nick' down each edge of the face of coal to be worked so as to loosen it.
nickings	The small coals made in nicking.
onsetter	The man who operates shaft signals in the pit-bottom, and pushes the tubs on to the cage.
opencast	Working coal by quarrying.
outbye	Away from the face.
outcrop	An exposed seam at the surface – usually the start of drift mining.
outstroke	A privilege permitting coals to be brought from one property and drawn to the surface at another.
overcast	Where air courses and roadways pass over each other.
overman	A supervisory official immediately below the under-manager.
pack	A stone wall built near the face to help take the weight of the earth.
panel	In bord-and-pillar, the workings are divided into districts (panels) *working* enclosed within barriers of coal so as to restrict the spread of explosions and to facilitate the working of pillars with a minimum of convergence of the strata.
parting	Horizontal break in a seam often filled with dirt or other foreign matter.
peas	Small coals.
peggy	In Durham, the handpick.
pick	The tool of the hewer for excavating his coal.

pillar	Mass of coal left to support the roof after excavation, *see* bord-and pillar
pills	The powder charges used in shot-firing.
pit	A single shaft; or, a pair of shafts (upcast and downcast); or, a colliery.
pit eye	The area around the bottom of the shaft.
pit heap	The slag or waste heap.
pit whip	Pick shaft carried by the putter.
place	The work place at the coal face where the collier won the coal. Each collier had his own place.
pony driver	A boy who had charge of a pony which pulled the tubs along the haulage roads.
pout/punch	Tool used by deputies in drawing timber out of a dangerous place.
pricker	Long thin rod with a flat hook on the end used by the shot-firer.
prop	Piece of wood or steel set vertically to support the roof. Horizontal bars were in later practice set up to the roof over the props.
prove	To confirm the existence or location of a seam of coal.
puffler	In Yorkshire, the leader of the working group.
punch-prop	A short prop.
putter	Young worker who brings the empty tubs to the face, and takes the full ones away, either with the help of a pony, or by hand.
rake	Rakes and forks (or screens) used by hewers to separate the small coal from the large.
ramble	Loose stone lying above the top of the coal.
rapper	A signal hammer at the top of the pit communicating with the bottom.
reckoning day	The day on which the miner receives his pay note.
refuge stall	A retreat used during the passing of trains.
regulator	A wooden partition with a shutter in an airway which restricts the flow of air inbye.
returning gallery	A tunnel through which air passes from the workings to the upcast shaft.
rib cutting	A tunnel cut through the coal, usually at right angles to a road.

ride	To travel in the cage up or down.
robbing	Partial excavation of pillars on a second working, *see* broken
rolley	A wheeled carriage formerly used in Northumberland and Durham to carry tubs or corves from the 'flat' to the pit-bottom.
rolley way	The main underground road.
rook	In the East Midlands, a capacity measure; stack of coal of a given size (say 2yds. x 1 yd. x 1 yd.). Hence stacking coal was called 'rocking'.
round coals	Large, best coal from which the small has been separated.
royalties	Rights of mineral ownership vested in the freehold, and rents accruing from exploitation of that mineral
scaffold	A platform built (usually of wood) across a shaft rendering the lower part inaccessible.
score	The number of tubs on which the putter's wage was calculated. Not necessarily twenty.
scoreprice	Pitmen's wages, the price current for filling a 'score'.
screens	The surface equipment where the coal is washed, sorted and graded.
seam	A strip or strata of coal.
seggar	Soft stone lying on coal-seams, used for making into bricks and coping-stones.
set	A train of tubs; or, a column of pumps.
set out	A corf or tub of coal confiscated for containing less than a stipulated amount.
setters	Large pieces of coal.
shaft	Vertical passage down into the mine from the surface.
shifter	In Northumberland and Durham, men paid so much per shift, elsewhere called 'daywagemen' or 'datallers'. These were usually employed on repair work.
shot firing	The blasting of coal or stone.
shotstick	A round stick on which a paper cartridge is rolled.
siddle	The inclination of a seam of coal.
sinker	A skilled man who contracted to sink new shafts.
sliding scale	A method providing for wages to vary automatically with the selling price of coal.

sliding-spears	Extend from top to bottom of the shaft for guiding the cages.
small coal	Coals of a certain size assessed after screening, or underground, and either wasted or sold cheaply, usually for export
smart money	Weekly payments to men off work through injury.
snap	In the Midlands, the mineworker's lunch.
sooty coal	Dull, soft coal.
sough/suff/ surf	Drainage adit or tunnel.
splint	Coarse, grey-looking coal
split	A branch in the ventilation system.
stable	*see* neuk.
staith/staithe	On the banks of the Tyne and the Wear, constructions for transferring coal into boats.
stall and pillar	Working by driving roadways ('stalls') into an area of coal, leaving pillars to support the roof.
standage	A place set apart for holding accumulations of water in the pit until pumped out.
staple	A small shaft running from one seam to another.
steam jets	Nozzles at the bottom of the upcast shaft through which steam was passed to cause convection currents for the ventilation. Considered safe since there were no naked flames to ignite inflammable gas.
steel mill	Before the invention of the safety lamp, used to provide light underground.
stenting	Narrow passage driven at right angles to the headways or drifts, for the purpose of ventilation.
stick	A strike.
stinkdamp or sulphurdamp	Sulphuretted hydrogen.
stint	A man's allotted task at the face.
stook	A small block or pillar left as a support.
stoop and room	*see* stall and pillar.
stopping	Air-tight door or barrier underground.

stover, stever or staver	In the East Midlands, the senior member of a butty partnership taking charge of the surface arrangements.
stythe	Choke-damp or black-damp.
sulphur	An old miner's word for fire-damp.
sump	The wall at the bottom of a shaft from which water may be pumped.
taker-off	Someone, usually a boy, who unhitched tubs from the endless rope.
tally stick	Wooden stick inserted in coal tub to indicate which man or group produced the coal.
ten	A capacity measure equal to about 52 tons used as a unit in mineral rent agreements in some districts. The rent is called a tentale rent.
thill	The floor of the seam or roadway.
thirl/thurl	To cut through a wall of coal; to join two roadways by cutting through.
thrusts	The crushing of coal pillars by pressure from the roof
thurst	*see* goaf.
timber leader	The modern equivalent of the hewer.
token	Disc bearing the putter or hewer's number and attached to the tub to identify it for the purpose of payment.
token-hanger	Boy who arranges the tokens attached to each corf to indicate the hewer of its contents.
tommy shop	Shop owned by a colliery proprietor or butty where pay tickets were exchanged for goods under the Truck System.
toom/tume	Empty.
top deck	The upper floor of the cage.
tram	Underground vehicle.
trammer	*see* putter.
tram plate	Flanged or angle rail on which trams with plain wheels ran.
trapper	Young (often very young) boy employed to open and close the air doors.
trimmer	Person who spreads the coals in the waggons or carriages in which the coals are conveyed along a railway from the top of the pit to the staith.
truck system	Payment of wages in kind.

tub	A small wagon used for conveying coal.
tubbing	Shaft lining of wood or iron to hold back the water.
under-looker	Official.
under-viewer	Deputy manager.
upbrow	A road driven uphill to the seam.
upcast	The ventilation shaft that carries the foul air away from the workings.
vend	Cartel of eighteenth century Northumberland and Durham coalowners.
ventilation door	A wooden door that directs the flow of fresh air round the workings of the mine.
viewer	In the early days of mining, the manager of the colliery.
wailer	Boy employed in wagons to pick stones missed during screening.
waste	The old and abandoned workings.
wasteman	Man who travels all the wastes or old workings, clears away falls of stone and attends especially to obstructions of the ventilation.
whim-gin	A horse-driven winding device.
whole	'Working in the whole' means working virgin coal by the bord-and pillar method; as distinct from 'working in the broken' where only the pillars are left to be got.
windy pick	Small pneumatic drill.
winning	The opening-up of a new pit, drift, face or gate.
wood and water leaders	Boys who carry props and wood to the various parts of the pit in which they are required.
workings	The excavations of a colliery.
yard stick	A walking stick carried by all mine officials, latterly as a symbol of authority.

1. Coalmining in Great Britain

1   *New Encyclopaedia Britannica, Micropaedia.* vol. 3. (Chicago: Encyclopaedia Britannica Inc., 1997). p. 408.

2   Durham County Environmental Education Curriculum Study Group. *Coal mining in County Durham.* (Durham: the Group, 1993). p. 120.

3   Shepherd, R *Ancient mining.* (London: Elsevier, 1993). pp. 395-6.

4   Page, W ed *Victoria History of the County of Somerset.* vol. 2. (London: Constable, 1911). p. 379.

5   Page, W ed *Victoria History of the County of Nottingham.* vol. 2. (London: Constable, 1910). p. 324.

6   Page, W ed *Victoria History of the County of Durham.* vol. 2. (London: Constable, 1908). p. 322.

7   Glennie, P D 'Industry and towns 1500-1730', in Dodgshon, R.A. and Butlin, R.A. *Historical geography of England and Wales.* 2nd. ed. (London: Academic Press, 1990). p. 210.

8   Page, W. ed *Victoria History of the County of Gloucester.* vol. 2. (London: Constable, 1907). p. 219.

9   Quoted in *Historical Review of coal mining.* (London: Fleetway Press, 1925). p. 7.

10   Jones, C 'Coal, gas and electricity', in Pope, R ed. *Atlas of British social and economic history since c. 1700.* (London, Routledge, 1989). p. 68.

11   Armstrong, J 'Transport and trade', in Pope, R ed. 1989. p. 102

12   Glennie, P D 1990. p. 210.

13   Royle, E *Modern Britain: a social history 1750-1985.* (London: E. Arnold, 1987). p. 34.

14   *ibid*. p. 34.

15   Lawton, R and Pooley, C G *Britain 1740-1950: an historical geography*. (London: E. Arnold, 1992). p. 67.

16   Armstrong, J 1989. p. 132.

17   Court, W H B *Coal*. History of the Second World War, United Kingdom Civil Series. (London: H.M.S.O., 1951). pp. 6, 8.

18   Kirby, M W *The British coalmining industry, 1870-1946: a political and economic history*. (London: Macmillan, 1977). p. 26.

19   Pearce, M and Stewart, G *British political history 1867-1995: democracy and decline*. 2nd ed. (London: Routledge, 1996). p. 259.

20   Royle, E 1987. p. 144.

21   Pearce, M and Stewart, G 1996. pp. 354-355.

22   Kirby, M W 1977. p. 170.

23   Quoted in *ibid*. p. 174.

24   *ibid*. p. 174.

25   Jones, C 1989. p. 73.

26   Griffin, A R 1977. p. 174.

27   Kirkup, M *Eyewitness: the Great Northern Coalfield*. (Newcastle upon Tyne: TUPS Books, 1999). p. 234.

28   Grimshaw, P N *Sunshine miners: opencast coalmining in Britain 1942-1992*. (Mansfield: British Coal Opencast, 1992).

29   Information from the UK Coal website.

30   Clarke, P *Hope and glory: Britain 1900-1990*. (London: Penguin, 1997). p. 333.

31   Jones, C 1989. p. 79.

32   Glyn, A and Machin, S 'Colliery closures and the decline of the UK coal industry', *British Journal of Industrial Relations*, vol. 35, no. 2, June 1997, pp. 197-214. p. 197.

33   Jones, C 1989. p. 81.

34   Campbell, A. *et al*, eds *Miners, unions and politics, 1910-47*. (Aldershot: Scolar Press, 1996). p. 1.

35   Information from the UK Coal website.

36   *The Times*, 18 April 2000. p. 13.

2. Why is he different?

1   Whatley, C A., "Scottish 'collier serfs', British coal workers? Aspects of Scottish collier society in the eighteenth century', *Labour History Review*, vol. 60, no. 2, Autumn 1995. p. 76.

2   Ashton, T S and Sykes, J *The coal industry of the eighteenth century*. 2nd. ed. (Manchester: Manchester University Press, 1964). p. 161.

3   Benson, J *British coalminers in the nineteenth century: a social history*. (Dublin: Gill and Macmillan, 1980). p. 93.

4   Jevons, H S *The British coal trade*. (Newton Abbot: David & Charles, 1969), first published 1915. p. 637.

5   Atkinson, G L *The miners' heritage: a history of the Durham Aged Mineworkers' Homes Association 1898-1997*. (Durham: County Durham Books, 1997).

6   Benson, J 1980. p. 106.

7   Pryce, W T R ed *From family history to community history*. (Cambridge: Cambridge University Press, 1994). p. 5.

8   Benson, J 1980. p. 126.

9   Jevons, H S 1969. p. 624.

10   Benson, J 1980. p. 127.

11   Lawson, J *Peter Lee*. (London: Hodder & Stoughton, 1936). pp. 7, 39

12   Quoted in Benson, J 1980. p. 126.

13   Benson, J 'Colliery disaster funds, 1860-1897', *International Review of Social History*, vol. xix, pt. 1, 1974, pp. 73-85. p. 75.

14   Snell, S 'Coal mining and the health of colliers', *Journal of the Sanitary Institute*, vol. xvi, 1895, pp. 105-24. p. 107.

15   Church, R *The history of the British coal industry, vol. 3., 1830-1913: Victorian pre-eminence*. (Oxford: Clarendon Press, 1986). p. 595.

16   Tonks, D 'A kind of life insurance: the coal-miners of North-East England 1860-1920', *Family and Community History*, vol. 2, no. 1, May 1999. pp. 46-58

17   Benson, 1980. p. 121.

18   Macfarlane, J 'Essay in oral history: Denaby Main – a South Yorkshire mining village – An interview with Robert Henry Shephard', *Bulletin of the Society for the Study of Labour History*, no. 25, Autumn 1972. pp. 85-88.

19   Leifchild, J R *Our coal and our coal-pits*. (London: F. Cass, 1968), first published 1853, 2nd. ed. 1856. p. 197.

20   *ibid*. p. 196.

21   Quoted in Ashton, T S and Sykes, J 1964. pp. 85-86.

22   Quoted in Colls, R *The pitmen of the northern coalfield: work, culture, and protest, 1790-1850*. (Manchester: Manchester University Press, 1987). p. 104.

23   Quoted in Mitchell, B R *Economic development of the British coal industry 1800-1914*. (Cambridge: Cambridge University Press, 1984). p. 354.

24   *ibid*. p. 171.

25   Quoted in Hammond, J L and Hammond, B *The town labourer 1760-1832: the new civilisation*. (Stroud: A. Sutton, 1995), first published 1917. p. 68.

26   Midland Mining Commission, *First Report: South Staffordshire, Appendix*. (London: H.M.S.O., 1843). p.44.

27 Quoted in Whatley, C A 1995. p. 67.

28 Quoted in Fynes, R *The miners of Northumberland and Durham.* (Newcastle upon Tyne: Davis Books, 1986), first published 1873. p. 145.

29 Children's Employment Commission, *Appendix to first report of Commissioners: Mines, part I: Reports and evidence from Sub-Commissioners.* (London: H.M.S.O., 1842). p. 590.

3. Work

1 Griffin, A R *Coalmining,* (London: Longman, 1971). pp. 2-3.

2 *ibid.* p. 49.

3 Hudson, M *Coming back brockens: a year in a mining village.* (London: J. Cape, 1994).

4 Griffin, A R *The British coalmining industry: retrospect and prospect.* (Buxton: Moorland Publishing, 1977). pp. 99-100.

5 Parkinson, G *True stories of Durham pit-life.* (London: C.H. Kelly, 1912). p.1.

6 Benson, J 1980. p. 48

7 Children's Employment Commission, *Appendix to first report of Commissioners: Mines, part I: Reports and evidence from Sub-Commissioners.* (London: H.M.S.O., 1842). p. 129.

8 Griffin, A R *Coalmining.* (London: Longman, 1971). pp. 115-129.

9 Quoted in Pollard, M *The hardest work under heaven: the life and death of the British coal miner.* (London: Hutchinson, 1984). p. 23.

10 Scott, W *An earnest address and urgent appeal to the people of England, in behalf of the oppressed and suffering pitmen, ….* (Newcastle upon Tyne, 1831). p. 5.

11 Jevons, H Stanley *The British coal trade.* (Newton Abbot: David & Charles, 1969), first published 1915. p. 365.

12 Anderson, D *Coal: a pictorial history of the British coal industry.* (Newton Abbot: David & Charles, 1982). p. 63.

13 Duckham, H and Duckham, B *Great pit disasters: Great Britain 1700 to the present day.* (Newton Abbot: David & Charles, 1973). p. 184.

14 Baxter, P J *et al* eds Hunter's *diseases of occupations,* 9th ed. (London: Arnold, 2000). pp. 682-686.

15 Anderson, D 1982. p. 65.

16 Winstanley, I *Mining deaths in Great Britain.* 7 vols. (Wigan: Picks Publishing, 1995).

17 Ritson, J A S 'The history of the development of mine rescue work in collieries of this country', in Mining Association of Great Britain *Historical review of coal mining.* (London: Fleetway Press, 1925), pp. 239-252.

18   Hayes, G *Coal mining*. (Princes Risborough: Shire Publications, 2000). p. 26.

19   Hodges, F 'The miner and his trade union', in Mining Association of Great Britain. 1925. pp. 334-350.

20   Colls, R *The pitmen of the northern coalfield: work, culture and protest, 1790-1850*. (Manchester: Manchester University Press, 1987). p. 30.

21   *A Voice from the Coal Mines*. (1825).

22   Griffin, A R 1977. p. 88.

23   Hodges, F 1925. p. 338.

24   Griffin, A R 1977. p. 88.

25   Mitchell, B R *Economic development of the British coal industry 1800-1914*. (Cambridge: Cambridge University Press, 1984). p. 179.

26   *ibid*. pp. 179-180.

27   Griffin, A R 1977. p. 91.

28   Hodges, F 1925. pp. 346.

29   Kirby, M W *The British coalmining industry, 1870-1946: a political and economic history*. (London: Macmillan, 1977). p. 47.

30   Hodges, F 1925. p. 349.

31   Anderson, D *Coal: a pictorial history of the British coal industry*. (Newton Abbot: David & Charles, 1982). p. 86.

32   Kirby, M W 1977. pp. 92-97.

33   Waller, R J *The Dukeries transformed: the social and political development of a twentieth century coalfield*. (Oxford: Clarendon Press, 1983). p. 108.

34   Kirby, M W 1977. pp. 179-181.

35   Griffin, A R 1977. p. 98.

36   Pearce, M and Stewart, G *British political history 1867-1995: democracy and decline*. 2nd edition. (London: Routledge, 1996). p. 541.

37   *ibid*. p. 542.

38   Church, R *et al* 'Towards a history of British miners' militancy', *Bulletin of the Society for the Study of Labour History*, vol. 54, part 1, Spring 1989, pp. 21-36. p. 21.

39   *Report of the Royal Commission on the Coal Industry (1925), vol. 1., Report*. (Cmd. 2600). (London: H.M.S.O., 1926. p. 108.

40   Quoted in Moyes, W A *The Banner book: a study of the banners of the lodges of the Durham Miners' Association*. (Newcastle upon Tyne: F. Graham, 1974). p. 7.

41   Quoted in Gorman, J *Banner bright: an illustrated history of trade union banners*. new ed. (Buckhurst Hill, Essex: Scorpion Publishing, 1986). p. 7.

42   Fynes, R 1986. p. 24.

43   Quoted in Emery, N *Banners of the Durham coalfield*. (Stroud: Sutton Publishing, 1998). p. 129.

4. How did he live

1 Griffin, A R *Coalmining.* (London: Longman, 1971). p. 132.

2 Anderson, D *Coal: a pictorial history of the British coal industry.* (Newton Abbot: David & Charles, 1982). p. 68.

3 Chaplin, S 'Durham mining villages", in Bulmer, M ed *Mining and social change.* (London: Croom Helm, 1978). p. 63.

4 Children's Employment Commission, *Appendix to first report of Commissioners: Mines, part I: Reports and evidence from Sub-Commissioners.* (London: H.M.S.O., 1842). p. 143.

5 Jevons, H S *The British coal trade.* (Newton Abbot: David & Charles, 1969), first published 1915. pp. 127-128.

6 Quoted in Waller, R J *The Dukeries transformed: the social and political development of a twentieth century coalfield.* (Oxford: Clarendon Press, 1983). p. 76.

7 Benson, J *British coalminers in the nineteenth century: a social history.* (Dublin: Gill and Macmillan, 1980). p. 85.

8 *ibid.* p. 74.

9 *ibid.* p. 91.

10 George, D L *Coal and power.* (London: Hodder & Stoughton, 1924).

11 Benson, J 1980. p. 85.

12 Anderson, D 1982 p. 74.

13 Priestley, J B *English journey.* (London: Heinemann, 1934). p. 336.

14 Jevons, H S 1969. p. 654.

15 Waller, R J 1983. p. 257.

16 Howitt, W *Visits to remarkable places ... chiefly in the Counties of Durham and Northumberland.* 2nd. series. (London: Longman, 1842). p. 87.

17 Ackers, P 'Review essay: life and death: mining history without a coal industry', *HSIR*, no. 1, March 1996, pp. 159-70. p. 167.

18 Chaplin, S 1978. p. 77.

19 Leifchild, J R *Our coal and our coal-pits; the people in them, and the scenes around them.* (London: Longman, 1853). pp. 197, 200.

20 'Life and labour in the coal-fields', *Cornhill Magazine*, vol. v., March 1862, pp. 343-353. p. 352.

21 Morning Chronicle, *Labour and the poor in England and Wales 1849-1851.* (London: F. Cass, 1983). p. 63.

22 Quoted in Robson, A H *The education of children engaged in industry in England 1833-1876.* (London: Kegan Paul, 1931).

23 Children's Employment Commission, 1842. p. 727.

24 Such as Sill, M 'The diary of Matthias Dunn, colliery viewer, 1831-1836', *Local Historian*, vol. 17, no. 7, 1985. pp. 418-424 – the original diary is in

Newcastle upon Tyne City Library; or Smith, E *A pitman's notebook: the diary of Edward Smith Houghton Colliery Viewer 1749-1751*. ed. T. Robertson. (Newcastle upon Tyne: F. Graham, 1970).

25　Storm-Clark, C 'The miners, 1870-1970: a test-case for oral history', *Victorian Studies*, September 1971. pp. 49-74.

26　no. 25, Autumn 1972; no. 26, Spring 1973; no. 31, Autumn 1975.

27　Douglass, D 'The Durham pitman', in Samuel, R ed *Miners, quarrymen and saltworkers*. (London: Routledge, 1977). p. 282.

28　Burnett, J *Useful toil: autobiographies of working people from the 1820s to the 1920s*. (London: A. Lane, 1974).

29　Redmayne, R A S *Men, mines and memories*. (London: Eyre & Spottiswoode, 1942).

30　Howard, W S 'Miners' autobiography: text and control', *Labour History Review*, vol. 60, no. 2, Autumn 1995. pp. 89-99.

31　Klaus, H G *The literature of labour: two hundred years of working-class writing*. (New York: St Martin's Press, 1985). pp. 81-84.

32　Austis, B and Austis, R *Diary of a working man, 1872-1873: Bill Williams in the Forest of Dean*. (Stroud: A. Sutton, 1995); Rymer, E A 'The martyrdom of the mine, or a sixty year struggle for life', *History Workshop Journal*, no. 1 Spring, no. 2 Autumn, 1976; Robinson, J *Tommy Turnbull: a miner's life*. (Newcastle upon Tyne, TUPS Books, 1996).

33　Such as: Bellamy, J M and Saville, J *Dictionary of labour biography*. 9 vols. (London: Macmillan, 1972-1993); Burnett, J *et al* eds *Autobiography of the working class: an annotated critical bibliography*. 3 vols. (Brighton: Harvester Press, 1984-1989); Matthews, W *British autobiographies: an annotated bibliography of British autobiographies published or written before 1951*. (Archon Books, 1968), first published 1955.

5. Was he Church or Chapel?

1　Quoted in Davies, R E *Methodism*. 2nd. ed. (Peterborough: Epworth Press, 1985). p. 57.

2　Quoted in Pollard, M *The hardest work under heaven: the life and death of the British coal miner*. (London: Hutchinson, 1984). p. 84.

3　Dexter, J *A guide to Radstock Museum*. (Radstock, Somerset: Radstock Museum, 1999). p. 19.

4　Pollard, M 1984. p. 85.

5　Milburn, G E *The travelling preacher: John Wesley in the North East 1742-1790*. (Wesley Historical Society (North East Branch), 1987). p. 33.

6　*ibid*. p. 36.

7   Wesley, J *The letters of John Wesley.* ed. J. Telford, vol. v. (London: Epworth Press, 1931). p. 121.

8   Patterson, W M *Northern Primitive Methodism.* (London: E. Dalton, 1909). p. 385.

9   Quoted in Williams, J E *The Derbyshire miners: a study in industrial and social history.* (London: Allen & Unwin, 1962). p. 76.

10 Petty, J *The history of the Primitive Methodist Connexion.* new ed. (London: J. Dickenson, 1880). p. 426.

11 Hammond, J L and Hammond, B *The town labourer 1760-1832: the new civilisation.* (London: Longman, 1917, reprinted Stroud: A. Sutton. 1995). p. 272.

12 Petty, J 1880. p. 180.

13 *Report of the Commissioner appointed ... to inquire into ... the state of the population in the mining districts, 1846.* (London: Her Majesty's Stationery Office, 1846). p. 14.

14 McLeod, H *Religion and society in England, 1850-1914.* (Basingstoke: Macmillan, 1996). pp. 35-36.

15 Ginswick, J ed *Labour and the poor in England and Wales 1849-1851, vol. II, Northumberland and Durham, Staffordshire, the Midlands.* (London: F. Cass, 1983). pp. 101-102.

16 Lambert, W R 'Some working-class attitudes towards organised religion in nineteenth-century Wales', in Parsons, G ed *Religion in Victorian Britain, vol. iv, Interpretations.* (Manchester: Manchester University Press, 1998). p. 97.

17 Hempton, D *The religion of the people: Methodism and popular religion c. 1750-1900.* (London: Routledge, 1996). p. 58.

18 Quoted in Benson, J *British coalminers in the nineteenth century: a social history.* (Dublin: Gill and Macmillan, 1980). p. 166.

19 Church, R *The history of the British coal industry, vol. 3: 1830-1913: Victorian pre-eminence.* (Oxford: Clarendon Press, 1985). p. 625.

20 Quoted in Duckham, B F *A history of the Scottish coal industry, vol. 1, 1700-1815: a social and industrial history.* (Newton Abbott: David & Charles, 1970). p. 289.

21 Benson, J 1980. p. 165.

22 Quoted in Neasham, G *North-country sketches: notes, essays and reviews.* (Durham: G. Neasham, 1893). p. 34.

23 Quoted in Williams, J E *The Derbyshire miners: a study in industrial and social history.* (London: Allen & Unwin, 1962). p. 466.

24 Quoted in Ramsey, A M *Miners and bishops.* City of Durham Annual Lecture, 11 November 1983. p.7.

25   Petty, J 1880. p. 179.

26   Hempton, D 1996. p. 58.

27   Andrews, S *Methodism and society*. (London: Longman, 1970). p. 82.

28   Wearmouth, R F *Methodism and the struggle of the working classes 1850-1900*. (Leicester: E. Backus, 1954). p. 173.

29   *Report, 1846.* 1846. p. 8.

6. How did he spend his leisure time?

1   Birch, C J *The development of working-class leisure in the coalmining regions of Staffordshire, 1840-1914*. 2003. (www.sulc.org/cjb/pdf/coalminers.doc).

2   Quoted in *ibid.* p. 2.

3   Jevons, H S *The British coal trade*. (Newton Abbot: David & Charles, 1969), first published 1915. p. 626.

4   Quoted in Benson, J *British coalminers in the nineteenth century: a social history*. (Dublin: Gill and Macmillan, 1980. p. 146.

5   Birch, C J 2003. p. 7.

6   Griffin, A R *The British coalmining industry: retrospect and prospect*. (Buxton: Moorland Publishing, 1977). p. 160.

7   Quoted in Benson, J 1980. p. 151.

8   Jevons, H S 1969. p. 628.

9   Quoted in Benson, J 1980. pp. 153-154.

10   Jevons, H S 1969. pp. 628-629.

11   Quoted in Emery, N *The coalminers of Durham*. (Stroud: Sutton Publishing, 1992). p. 179.

12   Children's Employment Commission, *Appendix to First Report of Commissioners: Mines Part I Reports and Evidence from Sub-Commissioners*. (London: H.M.S.O., 1842). p. 136.

13   Redmayne, R A S *Men, mines and memories*. (London: Eyre & Spottiswoode, 1942). pp. 8-9.

14   Birch, C J 2003. p. 17.

7. Was he a Bevin Boy?

1   Weiler, P *Ernest Bevin*, (Manchester: Manchester University Press, 1993). p. 118.

2   Quoted in Bullock, A *The life and times of Ernest Bevin, vol. 2, Minister of Labour 1940-1945*, (London: Heinemann, 1967). p. 257.

3   Court, W H B *Coal*, History of the Second World War United Kingdom Civil Series. (London: H.M.S.O., 1951). p. 304.

4   *Parliamentary Debates (Hansard), House of Commons official report, 24th November to 17th December, 1943*, (London: H.M.S.O., 1943), 2 December 1943, col. 521.

5   Bullock, A 1967. p. 260.

6   Pearce, M and Stewart, G *British political history 1867-1995: democracy and decline*. 2nd edition. (London: Routledge, 1996). p. 439.

8. But my mining ancestor was a woman

1   Pollard, M *The hardest work under heaven: the life and death of the British coal miner*. (London: Hutchinson, 1984). p. 62.

2   Quoted in *ibid*. pp. 64-65.

3   Children's Employment Commission, *First report of the Commissioners: Mines; Appendix to first report, parts I and II: Reports and evidence from Sub-Commissioners*. (London: H.M.S.O., 1842). p. 95.

4   Quoted in Pollard, M 1984. p. 72.

5   John, A V *By the sweat of their brow: women workers at Victorian coal mines*. (London: Croom Helm, 1980). p. 71.

6   *ibid*. p. 227.

7   Hudson, D *Munby: man of two worlds*. (London: J. Murray, 1972).

8   Hiley, M *Victorian working women: portraits from life*. (London: G. Fraser, 1979).

9. The arts

1   Galloway, R L *Annals of coal mining and the coal trade*. vol. 1 (first series, up to 1835), Newton Abbott: David & Charles, 1971), first published, 1898. pp. 5-6.

2   *Coal: British mining in art 1680-1980*. (London: Arts Council, [1982]). p. 7.

3   *ibid*. p. 12.

4   *ibid*. p. 20.

5   *ibid*. p. 26.

6   Feaver, W *Pitmen painters: the Ashington Group 1934-1984*. (London: Chatto & Windus, 1988). p. 9.

7   Quoted in *ibid*. p. 113.

8   *Coal*. 1982. p. 37

9   Klaus, H G *The literature of labour: two hundred years of working-class writing*. (New York: St Martin's Press, 1985). p. 62.

10   Croft, A 'Who was Harry Heslop?' in Heslop, H *Out of the old earth*. (Newcastle upon Tyne: Bloodaxe Books, 1994). p. 37.

11 Orwell, G *The road to Wigan Pier*. (London: Gollancz, 1937); Lawrence, D H 'Nottinghamshire and the mining countryside', in *Phoenix: the posthumous papers of D.H. Lawrence*. (London: Heinemann, 1936). pp. 133-140.

12 Over one hundred can be identified from such sources as *Cumulated Fiction Index* (7 vols, 1960-1996) and Baker, E A *A guide to the best fiction in English*. (London: Routledge, 1913).

13 Chaplin, S *A tree with rosy apples*. (Newcastle upon Tyne, 1972). p. 24.

14 Benson, J *British coalminers in the nineteenth century: a social history*. (Dublin: Gill and Macmillan, 1980). p. 158.

15 Colls, R, *The collier's rant: song and culture in the industrial village*. (London: Croom Helm, 1977). p. 18.

16 Klaus, H G 1985. p. 80.

17 Such as *Time Out film guide, 2009*. 17th ed. (London: Time Out Guides, 2008).

Appendix I. Glossary

1 Douglass, D 'Pit talk in county Durham', in Samuel, R ed *Miners, quarrymen and saltworkers*. (London: Routledge, 1977), pp. 297-348. p. 299.

INDEX

Aberdare Library   136

Apedale Heritage Centre   124

Archer, George   76

Arles   22

Armstrong, Tommy   70

art   67-68

Arts   67-70

Ashington Group   68

Astley Green Colliery Museum   106

autobiography   43-44

Ayrshire   12, 142-143

Ayrshire Archives   143

Bands and Banners   36

Bangor, University of Wales Library   133

banners   34-36

baptismal registers   16-17

Baptists   47-48

Barnsley Archives & Local Studies   110

Barrington, Shute   16

Beacon, The   101

Beamish – The North of England Open Air Museum   36, 97

beat knee   29

bell pit   6, 25, 156

Bell, Thomas   80

Bersham Colliery Mining Museum   133

Bevan, Nye   50

Bevin Boys   5, 57-60

Bevin Boys Association   59

Bevin, Ernest   57

bibliography of coalmining   149-152

big hewer   27

Big Pit Mining Museum   88, 137

Billy Elliot   70

Bilston   48

biography   43-44

Birmingham Archives and Heritage Service   123

Black Country   21-22, 38, 40, 121

Black Country Living Museum   124

Black Friday   4

black powder   72

Bolton Museum and Archive Service   103

Bond   20

bord and pillar   25

Brandt, Bill   68

Brickgarth   74

Bristol Record Office   127

Bristol University Library   127

British Coal   7

British Library   86

bronchitis   29

Bronze Age   1

Brotherton Library, University of Leeds   110
burial registers   17, 74
Burt, Thomas   32, 50
Butty   21-22
Calvinistic Methodists   48
Cannock Chase Mining Historical Society   125
Cefn Coed Colliery Museum   137
Census 1841   63
Census 1881   16, 73-74
Census 1911   10, 19
Chaplin, Sid   34, 38, 42, 69
Cheshire   103
Cheshire and Chester Archives and Local Studies Service   103
Chesterfield Library   117
Chicken, Edward   69
children   26
Children's Employment Commission Report, 1842   23, 24, 43, 55, 64, 68
Chowdene   46
Church of England   45-49
Church of Scotland   48
church records   16-17
coal   1
Coal Authority   7, 86
Coalmining History Resource Centre website   22, 23, 153

coal output   1, 3, 4-7
Coalbrookdale   68
Coleford   45
Coming Back Brokkens   iii-iv
communities   37-42
Congregationalists   47
Cook, A J   32
Cookson, Catherine   69
Coombes, B L   43
cooperative store   40
Cornwall   15
Corser, Haden   80
creel   62
Cronin, A J   69
Cumberland   3, 101-103
Cumbria   101-103
Cyfartha Castle Museum & Art Gallery   137
danger   29
Deal Library   131
Dean Heritage Centre   128
death   16, 22, 29-31, 34, 82
deputy   26
Derbyshire   21, 22, 116-121
Derbyshire Local Studies Library   118
Derbyshire Record Office   116
Dickens, Charles   69
disasters   22, 30, 71-84

diseases 16, 29
Doncaster Local Studies Library 111
drunkenness 54
Duckham, Baron 67
Dudley Archives and Local 121
  History Service
Dukeries 41
Dunfermline Central Library 144
Dunn, Matthias 43
Durham 2, 12, 13, 15, 16, 20, 32, 34,
  42, 48, 72, 92-101
Durham Aged Mineworkers' 13
  Homes Association
Durham, Archdeacon of 48
Durham Chronicle 76-78
Durham Clayport Library 95
Durham County Record Office 92
Durham Miners' Association 32
Durham Miners' Gala 34
Durham Mining Museum 11, 20, 97
Durham University Library 92
Easington Lane 72, 74, 78, 84
East Dunbartonshire Libraries 144
East Durham & Houghall 36
  Community College
East Kent Archives Centre 131
East Midlands 26, 46, 116-121
Eastwood 47
Edinburgh Central Library 145
Education Act 1870 54
Elemore Colliery 71-84
Elsecar Heritage Centre 112
Elliott family 70
Employment levels 4-7
Essential Work Order 57
eviction 12, 39
explorers 76
F Pit Museum, Washington 98
families 19, 42-43
family history 9-11
Fife 144-145, 147-148
film 70
Flintshire Record Office 132

Forest of Dean 2, 67
Franks, Robert 62
GENUKI 10, 154
gambling 56
gardening 55
Gateshead Library 95
General Strike, 1926 4, 33
Glamorgan Free Press 54
Glamorgan Record Office 134
Glasgow, Alex 69
Gloucestershire 15, 61, 127-131
Gloucestershire Archives 127
glossary 155-168
Goodchild, John Collection 109
Gwent Record Office 135
Haig Colliery 64, 102
Hair, Thomas 68
Hepburn, Tommy 32
Heslop, Harold 69
Hetton, Eppleton and Elemore
  Collieries Accident Fund 78
hewer 26-27
hewing putter 26
Hiley, Michael 64
Hipps, Margaret 62
History Shop, The Wigan 106
Hodges, Frank 50
Holloway, Ted 59
House of Commons Parliamentary 24
  Papers website
housing 12-13, 37-42
Howitt, William 42
Hudson, Mark iii
Hull History Centre 109
Imperial War Museum 59
in-breeding 19-20
Independents 47-48
Industrial Revolution 3
inquests 71
Inspectors of Mines 24, 29, 30, 71,
  80-82
Internet 152-154
Iron Mountain 86

Ironbridge Gorge Museums 125
Jack, Alison 62
Jevons, H S 12
Johnson, Thomas 72
Jones, Janet 40
Jordan, Thomas 43
Jude, Martin 32
Keele University Library 122
Kent 4, 15, 131
Kidwelly Industrial Museum 138
King, John Workshop Museum 119
Kingswood 45
Kirkcaldy 68
Kirkcaldy Museum and Art 145
  Gallery
Lanarkshire 13, 32, 39
Lancashire 3, 20, 22, 38, 46, 48, 61,
  63, 103-109
Lancashire Record Office/ 104
  Local Studies Library
Lawrence, D H 69
Lawson, Jack 15, 50
Lea Green Colliery 22
Lee, Peter 15, 41
Leicestershire 61, 119
Leifchild, John Roby 19-20, 42
leisure 53-56
libraries 54
literature 69
Liverpool Record Office 104
living conditions 37-44
Llewellyn, Richard 69
Loanhead Colliery 62
lodgers 13
London 1, 3
longwall 26
Lothians 12
Lovekin, Emanuel 43
Luke, John 72
Macdonald, Alexander 50
Maerdy 54
mailing lists 154
Marxism 50

Merthyr 39
Methodism 45-51
migration 13-16
Miners' Federation of 32, 50
  Great Britain
Miners' Institute 54
Miners' strike, 1984-85 6, 33, 70
Mines and Collieries Act 1842 64
Mines rescue 30
Mining Industry Act 1920 55
Modern Records Centre, 123
  University of Warwick
Moore, Henry 68
Morrison, Herbert 49
Munby, Arthur 64
Museum of Cannock Chase 125
Museum of Local Crafts and 107
  Industries, Burnley
Museum of South Yorkshire Life 112
Museum of Welsh Life 138
music 70
National Archives of Scotland 86, 143
National Archives, The 59, 85
National Coal Board 5-7, 33, 86
National Coal Mining 36, 87, 112
  Museum for England
National Library of Scotland 87, 144
National Library of Wales 87
National Mining Memorabilia 40
  Association
National Museum of 107
  Labour History
National Union of 33, 87, 121, 144,
  Mineworkers 151
Nationalisation 5
NEEMARC 93
Newcastle Library 96
Newland Church 2, 67
Newport Library 136
newspapers 74, 76, 86
North East England 3, 12, 13, 16, 20,
  25-26, 31, 32, 34, 37, 40, 43, 46, 50,
  70, 92-101

North East England Mining 93
Archive and Research Centre
North of England Institute of 96
Mining and Mechanical Engineers
North Wales 21, 61, 132-134
Northumberland 1, 12, 16-17, 20, 25,
32, 34, 37, 46, 58, 61, 68, 92-101
Northumberland Miners' Picnic 34
Nottingham 2
Nottingham University Library 117
Nottinghamshire 3, 21, 33, 41, 47,
116-121
Nottinghamshire Archives 117
Nottingham Local Studies Library 118
novels 69
nystagmus 29
opencast mining 5, 7
Orwell, George 69
overman 26, 47
Owen, Daniel Museum & Heritage 134
Centre
Palmer, A N Centre for Local 133
Studies and Archives
Palmer's Index to *The Times* 78
Parker, Henry Perlee 68
Parkinson, George 26
Paterson family 72-74
People's History Museum 88, 107
Pembrokeshire Record Office 135
Pickard, Ben 50
pillar and stall 6, 25
pit brow lasses 64
Plater, Alan 69
plays 69
pneumoconiosis 16, 29
poetry 69-70
Pontypool Museum 138
pony driver 26
Priestley, J B 41
Primitive Methodism 46, 47, 50
privatisation 7
public house 22, 54
Public Libraries Act 1850 54

Putter 26
Radstock Museum 129
Ranters 46
Relief fund 22, 71, 78
religion 45-51
Reports on the Circumstances 80-82
attending an explosion which
occurred at the Elemore Colliery on
the 2nd of December 1886. 1887
rescue stations 30
Rhondda Heritage Park 139
Rhondda Museum 139
RJB Mining 7
Roberts, W P 20
Robinson Library, Newcastle 94
University
Roman Catholics 17, 48
Rymer, Edward 44
St. Helens Local History and 105
Archives Library
St. Michael's Church, Lyons 74-78, 84
Salford Local History Library 105
and City Archives Centre
Samuel Commission 32, 34
Sankey Commission 32
Science Museum 7
Scotland 5, 12, 20, 21, 22-23, 37,
39, 40, 48, 61, 63, 86-87, 142-148
Scotsman, The 48
Scottish Mining Museum 63, 88, 145
sea coal 3
serfdom 20, 22-23, 48
Shepton Mallet 45
Shinwell, Manny 5
Shropshire, 21, 125-126
Silk Mill, Derby 118
Simpson, William 23
Skipsey, Joseph 70
slype 61-62
Snibston 119
Society of Genealogists 89
Somerset 1, 3, 15, 45, 61, 127-130

Somerset Archive & Record 128
  Service
song 27, 70
South Wales 3, 4, 12, 13, 15, 21, 32,
  36, 40-41, 46, 48, 54, 64, 134-142
South Wales Coalfield 36, 135
  Collection
South Wales Miners' Library 36, 135
South Wales Miners' Museum 139
Southwell, Bishop of 48
South Yorkshire 12, 38, 40, 41
sport 53-55
Staffordshire 5, 46, 59, 61, 86, 121-127
Staffordshire Record Office 122
Staffordshire University Library 123
Stockton and Darlington Railway 3
Stoke-on-Trent City Archives 122
Storey, David 69
strikes 4, 6, 20, 21, 32, 33-35, 50, 56
Summerlee - Museum of 146
  Scottish Industrial Life
Sunderland Daily Echo 76, 83
Sunderland Library 96
Sunderland Museum 98
Sunderland University 93
Sutherland, Graham 68
Swansea University Archives 136
Symons, Jelinger C 63
Taylor, Warwick H 59
Thomas, George 50
Thompson, George 74
Times, The 74, 78-80
Times Archive, The 78
TNA 59, 85
 trapper boy 26
Tremenheere, H S 55
Truck 21-22, 39-40
Turnbull, Tommy 44
Tyne and Wear Archives 93
Union of Democratic Miners 33
unions 31-34
Van Gogh 68
Vesting Day 5

Viewer 27, 47
wages 6, 20, 21, 32, 54
waggonways 3
Wakefield Learning and Local 111
  Studies Library
Wakefield Museum 113
Wales 1, 3, 4, 12-13, 15, 20, 21,
  32, 36, 40, 41, 46, 48, 54, 61, 64, 87-
  88, 132-142
Wardle, William 29
Warwick University Library 123
Warwickshire 46, 61
Washington F Pit Museum 98
web rings 154
websites 152-154
Welfare Scheme 55
Wesley, John 45-46, 50
Wesleyan Methodism 46, 47, 50, 78
West Midlands 3, 15, 46, 121-127
West Yorkshire Archives Service 110
Westhoughton Library 106
Wheldale Colliery 68
Whickham 2
Wigan 64, 69, 106
Williams, Bill 44
Willis, Norman 34
Wilson, John 50
Wilson, Thomas 69
Winding House 139
women 61-65
Women and the Pits website 23
Woodhorn 36, 94
work 25-36, 53, 61-65
Working Class Movement 105
  Library
World War I 4
World War II 5
Wrexham Local Studies 133
  and Archives
Yorkshire 12, 15, 21, 26, 32, 37-38,
  40, 41, 46, 63, 109-116
Young, Arthur 3

Founded in 1911 the Society of Genealogists (SoG) is Britain's premier family history organisation. The Society maintains a splendid genealogical library and education centre in Clerkenwell.

The Society's collections are particularly valuable for research before the start of civil registration of births marriages and deaths in 1837 but there is plenty for the beginner too. Anyone starting their family history can book free help sessions in the open community access area where free help can be given in searching online census indexes or looking for entries in birth, death and marriage indexes.

The Library contains Britain's largest collection of parish register copies, indexes and transcripts and many nonconformist registers. Most cover the period from the sixteenth century to 1837. Along with registers, the library holds local histories, copies of churchyard gravestone inscriptions, poll books, trade directories, census indexes and a wealth of information about the parishes where our ancestors lived.

Unique indexes include Boyd's Marriage Index with more than 7 million names compiled from 4300 churches between 1538-1837 and the Bernau Index with references to 4.5 million names in Chancery and other court proceedings. Also available are indexes of wills and marriage licences, and of apprentices and masters (1710-1774). Over the years the Society has rescued and made available records discarded by government departments and institutions but of great interest to family historians. These include records from the Bank of England, Trinity House and information on Teachers and Civil Servants.

Boyd's and other unique databases are published on line on www.origins.com, on www.findmypast.com and on the Society's own website www.sog.org.uk . There is free access to these and many other genealogical sites within the Library's Internet suite.

The Society is the ideal place to discover if a family history has already been researched with its huge collection of unique manuscript notes, extensive collections of past research and printed and unpublished family histories. If you expect to be carrying out family history research in the British Isles then membership is very worthwhile although non-members can use the library for a small search fee.

The Society of Genealogists is an educational charity. It holds study days, lectures, tutorials and evening classes and speakers from the Society regularly speak to groups around the country. The SoG runs workshops demonstrating computer programs of use to family historians. A diary of events and booking forms are available from the Society on 020 7553 3290 or on the website www.sog.org.uk .

Members enjoy free access to the Library, certain borrowing rights, free copies of the quarterly *Genealogists Magazine* and various discounts of publications, courses, postal searches along with free access to data on the members' area of our website and each quarter to our data on www.origins.com.

More details about the Society can be found on its extensive website at www.sog.org.uk

For a free Membership Pack contact the Society at:

14 Charterhouse Buildings,
Goswell Road,
London EC1M 7BA
Telephone 020 7553 3291
Fax 020 7250 1800

The Society is always happy to help with enquiries and the following contacts may be of assistance.

Library & shop hours:

Monday	Closed
Tuesday	10am – 6pm
Wednesday	10am – 6pm
Thursday	10am – 8pm
Friday	Closed
Saturday	10am – 6pm
Sunday	Closed

Contacts:

Membership
Tel: 020 7553 3291
Email: membership@sog.org.uk

Lectures & courses
Tel: 020 7553 3290
Email: events@sog.org.uk

Family history advice line
Tel: 020 7490 8911
See website for availability